Thomas Price

Tables of the value of gold and silver per ounce troy

At different degrees of fineness

Thomas Price

Tables of the value of gold and silver per ounce troy
At different degrees of fineness

ISBN/EAN: 9783337113506

Printed in Europe, USA, Canada, Australia, Japan

Cover: Foto ©Andreas Hilbeck / pixelio.de

More available books at **www.hansebooks.com**

TABLES

OF THE

VALUE OF GOLD AND SILVER

PER OUNCE TROY,

AT DIFFERENT DEGREES OF FINENESS.

WITH

OTHER TABLES, WHICH WILL BE FOUND USEFUL TO BANKERS, ASSAYERS, CHEMISTS, MERCHANTS, DEALERS IN BULLION AND ORES.

PREPARED BY

THOMAS PRICE

BULLION ROOMS, ASSAY OFFICE, CHEMICAL LABORATORY AND ORE FLOORS

524 SACRAMENTO STREET, SAN FRANCISCO.

A. J. LEARY, STATIONER AND PRINTER, Nos. 402 & 404 SANSOME STREET.
1880.

EXPLANATION OF GOLD TABLE.

The values per oz. of gold in the following tables, are computed from the simple *formula* that 387 ozs. of *pure* gold (1000 fine) are worth $8000. Hence, 1 oz. is worth $20.671834625, and the $\frac{1}{1000}$ of an oz. (decimally expressed as .001 fine) is worth $0.020671834625.

What we usually call *fineness*, therefore, is simply the *weight* of *fine* metal contained in any given quantity of mixed metals or alloys. For instance, in a gold or silver bar which is reported to be 850 fine, it is simply meant that in 1000 parts *by weight*, 850 are *fine* gold or *fine* silver, as the case may be.

In our mints, the value of gold is computed from *standard* weight; that is, gold which is 900 fine, that being the *fineness* of our gold coin as required by law. The *formula* in this case is, 43 ozs. of *standard* gold are worth $800. Hence, multiply standard ozs. by 800, and divide by 43, and you obtain the value.

EXAMPLE:—Take 123$\frac{13}{100}$ ozs. at 843 fine, and we obtain the result thus:

```
                  123.13   ozs. gross weight.
                    843    fineness of gold.
                  ───────
                  369 39
                  4 925 2
                  98 504
                  ───────
U. S. standard, 900 ) 103,798.59   ozs. of fine gold.
                  ───────
                  115.331   ozs. of st'd gold.
                    800
                  ───────
             43 ) 92,264.800
                  ───────
                  $2,145.69, value.
```

To find value per oz., divide total value ($2,145.69) by standard ozs. (115.331) and you have $18.60^{44} which will be seen by reference to the table is the value of 1 oz. of gold at 900 fine.

The value in this case would have been ascertained thus: By reference to the Gold Table and opposite .843, the value of one oz. .843 fine is $17.42 64. Hence,

```
                  123.13
                  174264
                  ───────
                  492 52
                  7 387 8
                  24 626
                  492 52
                  8 619 1
                  12 313
                  ───────
                  $2,145.69 value.
```

TABLE OF THE VALUE OF
GOLD
Per OUNCE TROY, AT DIFFERENT DEGREES OF FINENESS
–BY–
THOMAS PRICE.

FINE	DOLLARS	CENTS	FINE	DOLLARS	CENTS	FINE	DOLLARS	CENTS	FINE	DOLLARS	CENTS	FINE	DOLLARS	CENTS
0		00.00	10		20.67	20		41.34	30		62.02	40		82.69
½		01.03	½		21.71	½		42.38	½		63.05	½		83.72
1		02.07	11		22.74	21		43.41	31		64.08	41		84.75
½		03.10	½		23.77	½		44.44	½		65.12	½		85.79
2		04.13	12		24.81	22		45.48	32		66.15	42		86.82
½		05.17	½		25.84	½		46.51	½		67.18	½		87.86
3		06.20	13		26.87	23		47.55	33		68.22	43		88.89
½		07.24	½		27.91	½		48.58	½		69.25	½		89.92
4		08.27	14		28.94	24		49.61	34		70.28	44		90.96
½		09.30	½		29.97	½		50.65	½		71.32	½		91.99
5		10.34	15		31.01	25		51.68	35		72.35	45		93.02
½		11.37	½		32.04	½		52.71	½		73.39	½		94.06
6		12.40	16		33.07	26		53.75	36		74.42	46		95.09
½		13.44	½		34.11	½		54.78	½		75.45	½		96.12
7		14.47	17		35.14	27		55.81	37		76.49	47		97.16
½		15.50	½		36.18	½		56.85	½		77.52	½		98.19
8		16.51	18		37.21	28		57.88	38		78.55	48		99.22
½		17.57	½		38.24	½		58.91	½		79.59	½	1	00.26
9		18.60	19		39.28	29		59.95	39		80.62	49	1	01.29
½		19.64	½		40.31	½		60.98	½		81.65	½	1	02.33

BULLION ROOMS, ASSAY OFFICE, CHEMICAL LABORATORY AND ORE FLOORS,
524 Sacramento Street, San Francisco.

TABLE OF THE VALUE OF
GOLD
Per OUNCE TROY, AT DIFFERENT DEGREES OF FINENESS
–BY–
THOMAS PRICE.

FINE	DOLLARS	CENTS	FINE	DOLLARS	CENTS	FINE	DOLLARS	CENTS	FINE	DOLLARS	CENTS	FINE	DOLLARS	CENTS
50	1	03.36	60	1	24.03	70	1	44.70	80	1	65.37	90	1	86.05
½	1	04.39	½	1	25.06	½	1	45.74	½	1	66.41	½	1	87.08
51	1	05.43	61	1	26.10	71	1	46.77	81	1	67.44	91	1	88.11
½	1	06.46	½	1	27.13	½	1	47.80	½	1	68.48	½	1	89.15
52	1	07.49	62	1	28.17	72	1	48.84	82	1	69.51	92	1	90.18
½	1	08.53	½	1	29.20	½	1	49.87	½	1	70.54	½	1	91.21
53	1	09.56	63	1	30.23	73	1	50.90	83	1	71.58	93	1	92.25
½	1	10.59	½	1	31.27	½	1	51.94	½	1	72.61	½	1	93.28
54	1	11.63	64	1	32.30	74	1	52.97	84	1	73.64	94	1	94.32
½	1	12.66	½	1	33.33	½	1	54.01	½	1	74.68	½	1	95.35
55	1	13.70	65	1	34.37	75	1	55.04	85	1	75.71	95	1	96.38
½	1	14.73	½	1	35.40	½	1	56.07	½	1	76.74	½	1	97.42
56	1	15.76	66	1	36.43	76	1	57.11	86	1	77.78	96	1	98.45
½	1	16.80	½	1	37.47	½	1	58.14	½	1	78.81	½	1	99.48
57	1	17.83	67	1	38.50	77	1	59.17	87	1	79.84	97	2	00.52
½	1	18.86	½	1	39.53	½	1	60.21	½	1	80.88	½	2	01.55
58	1	19.90	68	1	40.57	78	1	61.24	88	1	81.91	98	2	02.58
½	1	20.93	½	1	41.60	½	1	62.27	½	1	82.95	½	2	03.62
59	1	21.96	69	1	42.64	79	1	63.31	89	1	83.98	99	2	04.65
½	1	23.00	½	1	43.67	½	1	64.34	½	1	85.01	½	2	05.68

BULLION ROOMS, ASSAY OFFICE, CHEMICAL LABORATORY AND ORE FLOORS,
524 Sacramento Street, San Francisco.

TABLE OF THE VALUE OF
GOLD

Per OUNCE TROY, AT DIFFERENT DEGREES OF FINENESS

-BY-

THOMAS PRICE.

FINE.	DOLLARS.	CENTS.	FINE.	DOLLARS.	CENTS.	FINE.	DOLLARS.	CENTS.	FINE.	DOLLARS.	CENTS.	FINE.	DOLLARS.	CENTS.
100	2	06.72	110	2	27.39	120	2	48.06	130	2	68.73	140	2	89.41
½	2	07.75	½	2	28.42	½	2	49.10	½	2	69.77	½	2	90.44
101	2	08.79	111	2	29.46	121	2	50.13	131	2	70.80	141	2	91.47
½	2	09.82	½	2	30.49	½	2	51.16	½	2	71.83	½	2	92.51
102	2	10.85	112	2	31.52	122	2	52.20	132	2	72.87	142	2	93.54
½	2	11.89	½	2	32.56	½	2	53.23	½	2	73.90	½	2	94.57
103	2	12.92	113	2	33.59	123	2	54.26	133	2	74.94	143	2	95.61
½	2	13.95	½	2	34.63	½	2	55.30	½	2	75.97	½	2	96.64
104	2	14.99	114	2	35.66	124	2	56.33	134	2	77.00	144	2	97.67
½	2	16.02	½	2	36.69	½	2	57.36	½	2	78.04	½	2	98.71
105	2	17.05	115	2	37.73	125	2	58.40	135	2	79.07	145	2	99.74
½	2	18.09	½	2	38.76	½	2	59.43	½	2	80.10	½	3	00.78
106	2	19.12	116	2	39.79	126	2	60.46	136	2	81.14	146	3	01.81
½	2	20.16	½	2	40.83	½	2	61.50	½	2	82.17	½	3	02.84
107	2	21.19	117	2	41.86	127	2	62.53	137	2	83.20	147	3	03.88
½	2	22.22	½	2	42.89	½	2	63.57	½	2	84.24	½	3	04.91
108	2	23.26	118	2	43.93	128	2	64.60	138	2	85.27	148	3	05.94
½	2	24.29	½	2	44.96	½	2	65.63	½	2	86.30	½	3	06.98
109	2	25.32	119	2	45.99	129	2	66.67	139	2	87.34	149	3	08.01
½	2	26.36	½	2	47.03	½	2	67.70	½	2	88.37	½	3	09.04

BULLION ROOMS, ASSAY OFFICE, CHEMICAL LABORATORY AND ORE FLOORS,
524 Sacramento Street, San Francisco.

TABLE OF THE VALUE OF
GOLD

Per OUNCE TROY, AT DIFFERENT DEGREES OF FINENESS

–BY–

THOMAS PRICE.

FINE	DOLLARS	CENTS	FINE	DOLLARS	CENTS	FINE	DOLLARS	CENTS	FINE	DOLLARS	CENTS	FINE	DOLLARS	CENTS
150	3	10.08	160	3	30.75	170	3	51.42	180	3	72.09	190	3	92.76
½	3	11.11	½	3	31.78	½	3	52.45	½	3	73.13	½	3	93.80
151	3	12.14	161	3	32.82	171	3	53.49	181	3	74.16	191	3	94.83
½	3	13.18	½	3	33.85	½	3	54.52	½	3	75.19	½	3	95.87
152	3	14.21	162	3	34.88	172	3	55.56	182	3	76.23	192	3	96.90
½	3	15.25	½	3	35.92	½	3	56.59	½	3	77.26	½	3	97.93
153	3	16.28	163	3	36.95	173	3	57.62	183	3	78.29	193	3	98.97
½	3	17.31	½	3	37.98	½	3	58.66	½	3	79.33	½	4	00.00
154	3	18.35	164	3	39.02	174	3	59.69	184	3	80.36	194	4	01.03
½	3	19.38	½	3	40.05	½	3	60.72	½	3	81.40	½	4	02.07
155	3	20.41	165	3	41.09	175	3	61.76	185	3	82.43	195	4	03.10
½	3	21.45	½	3	42.12	½	3	62.79	½	3	83.46	½	4	04.13
156	3	22.48	166	3	43.15	176	3	63.82	186	3	84.50	196	4	05.17
½	3	23.51	½	3	44.19	½	3	64.86	½	3	85.53	½	4	06.20
157	3	24.55	167	3	45.22	177	3	65.89	187	3	86.56	197	4	07.24
½	3	25.58	½	3	46.25	½	3	66.93	½	3	87.60	½	4	08.27
158	3	26.61	168	3	47.29	178	3	67.96	188	3	88.63	198	4	09.30
½	3	27.65	½	3	48.32	½	3	68.99	½	3	89.66	½	4	10.34
159	3	28.68	169	3	49.35	179	3	70.03	189	3	90.70	199	4	11.37
½	3	29.72	½	3	50.39	½	3	71.06	½	3	91.73	½	4	12.40

Bullion Rooms, Assay Office, Chemical Laboratory and Ore Floors,

524 Sacramento Street, San Francisco.

TABLE OF THE VALUE OF

GOLD

Per OUNCE TROY, AT DIFFERENT DEGREES OF FINENESS

–BY–

THOMAS PRICE.

FINE.	DOLLARS.	CENTS.	FINE.	DOLLARS.	CENTS.	FINE.	DOLLARS.	CENTS.	FINE.	DOLLARS.	CENTS.	FINE.	DOLLARS.	CENTS.
200	4	13.44	**210**	4	34.11	**220**	4	54.78	**230**	4	75.45	**240**	4	96.12
½	4	14.47	½	4	35.14	½	4	55.81	½	4	76.49	½	4	97.16
201	4	15.50	211	4	36.18	221	4	56.85	231	4	77.52	241	4	98.19
½	4	16.54	½	4	37.21	½	4	57.88	½	4	78.55	½	4	99.22
202	4	17.57	212	4	38.24	222	4	58.91	232	4	79.59	242	5	00.26
½	4	18.60	½	4	39.28	½	4	59.95	½	4	80.62	½	5	01.29
203	4	19.64	213	4	40.31	223	4	60.98	233	4	81.65	243	5	02.33
½	4	20.67	½	4	41.34	½	4	62.02	½	4	82.69	½	5	03.36
204	4	21.71	214	4	42.38	224	4	63.05	234	4	83.72	244	5	04.39
½	4	22.74	½	4	43.41	½	4	64.08	½	4	84.75	½	5	05.43
205	4	23.77	215	4	44.44	225	4	65.12	235	4	85.79	245	5	06.46
½	4	24.81	½	4	45.48	½	4	66.15	½	4	86.82	½	5	07.49
206	4	25.84	216	4	46.51	226	4	67.18	236	4	87.86	246	5	08.53
½	4	26.87	½	4	47.55	½	4	68.22	½	4	88.89	½	5	09.56
207	4	27.91	217	4	48.58	227	4	69.25	237	4	89.92	247	5	10.59
½	4	28.94	½	4	49.61	½	4	70.28	½	4	90.96	½	5	11.63
208	4	29.97	218	4	50.65	228	4	71.32	238	4	91.99	248	5	12.66
½	4	31.01	½	4	51.68	½	4	72.35	½	4	93.02	½	5	13.70
209	4	32.04	219	4	52.71	229	4	73.39	239	4	94.06	249	5	14.73
½	4	33.07	½	4	53.75	½	4	74.42	½	4	95.09	½	5	15.76

BULLION ROOMS, ASSAY OFFICE, CHEMICAL LABORATORY AND ORE FLOORS,

524 Sacramento Street, San Francisco.

TABLE OF THE VALUE OF
GOLD
Per OUNCE TROY, AT DIFFERENT DEGREES OF FINENESS

-BY-

THOMAS PRICE.

FINE	DOLLARS	CENTS	FINE	DOLLARS	CENTS	FINE	DOLLARS	CENTS	FINE	DOLLARS	CENTS	FINE	DOLLARS	CENTS
250	5	16.80	**260**	5	37.47	**270**	5	58.14	**280**	5	78.81	**290**	5	99.48
½	5	17.83	½	5	38.50	½	5	59.17	½	5	79.84	½	6	00.52
251	5	18.86	261	5	39.53	271	5	60.21	281	5	80.88	291	6	01.55
½	5	19.90	½	5	40.57	½	5	61.24	½	5	81.91	½	6	02.58
252	5	20.93	262	5	41.60	272	5	62.27	282	5	82.95	292	6	03.62
½	5	21.96	½	5	42.64	½	5	63.31	½	5	83.98	½	6	04.65
253	5	23.00	263	5	43.67	273	5	64.34	283	5	85.01	293	6	05.68
½	5	24.03	½	5	44.70	½	5	65.37	½	5	86.05	½	6	06.72
254	5	25.06	264	5	45.74	274	5	66.41	284	5	87.08	294	6	07.75
½	5	26.10	½	5	46.77	½	5	67.44	½	5	88.11	½	6	08.79
255	5	27.13	265	5	47.80	275	5	68.48	285	5	89.15	295	6	09.82
½	5	28.17	½	5	48.84	½	5	69.51	½	5	90.18	½	6	10.85
256	5	29.20	266	5	49.87	276	5	70.54	286	5	91.21	296	6	11.89
½	5	30.23	½	5	50.90	½	5	71.58	½	5	92.25	½	6	12.92
257	5	31.27	267	5	51.94	277	5	72.61	287	5	93.28	297	6	13.95
½	5	32.30	½	5	52.97	½	5	73.64	½	5	94.32	½	6	14.99
258	5	33.33	268	5	54.01	278	5	74.68	288	5	95.35	298	6	16.02
½	5	34.37	½	5	55.04	½	5	75.71	½	5	96.38	½	6	17.05
259	5	35.40	269	5	56.07	279	5	76.74	289	5	97.42	299	6	18.09
½	5	36.43	½	5	57.11	½	5	77.78	½	5	98.45	½	6	19.12

BULLION ROOMS, ASSAY OFFICE, CHEMICAL LABORATORY AND ORE FLOORS,
524 Sacramento Street, San Francisco.

TABLE OF THE VALUE OF
GOLD
Per OUNCE TROY, AT DIFFERENT DEGREES OF FINENESS
–BY–
THOMAS PRICE.

FINE	DOLLARS	CENTS	FINE	DOLLARS	CENTS	FINE	DOLLARS	CENTS	FINE	DOLLARS	CENTS	FINE	DOLLARS	CENTS
300	6	20.16	**310**	6	40.83	**320**	6	61.50	**330**	6	82.17	**340**	7	02.84
½	6	21.19	½	6	41.86	½	6	62.53	½	6	83.20	½	7	03.88
301	6	22.22	311	6	42.89	321	6	63.57	331	6	84.24	341	7	04.91
½	6	23.26	½	6	43.93	½	6	64.60	½	6	85.27	½	7	05.94
302	6	24.29	312	6	44.96	322	6	65.63	332	6	86.30	342	7	06.98
½	6	25.32	½	6	45.99	½	6	66.67	½	6	87.34	½	7	08.01
303	6	26.36	313	6	47.03	323	6	67.70	333	6	88.37	343	7	09.04
½	6	27.39	½	6	48.06	½	6	68.73	½	6	89.41	½	7	10.08
304	6	28.42	314	6	49.10	324	6	69.77	334	6	90.44	344	7	11.11
½	6	29.46	½	6	50.13	½	6	70.80	½	6	91.47	½	7	12.14
305	6	30.49	315	6	51.16	325	6	71.83	335	6	92.51	345	7	13.18
½	6	31.52	½	6	52.20	½	6	72.87	½	6	93.54	½	7	14.21
306	6	32.56	316	6	53.23	326	6	73.90	336	6	94.57	346	7	15.25
½	6	33.59	½	6	54.26	½	6	74.94	½	6	95.61	½	7	16.28
307	6	34.63	317	6	55.30	327	6	75.97	337	6	96.64	347	7	17.31
½	6	35.66	½	6	56.33	½	6	77.00	½	6	97.67	½	7	18.35
308	6	36.69	318	6	57.36	328	6	78.01	338	6	98.71	348	7	19.38
½	6	37.73	½	6	58.40	½	6	79.07	½	6	99.74	½	7	20.41
309	6	38.76	319	6	59.43	329	6	80.10	339	7	00.78	349	7	21.45
½	6	39.79	½	6	60.47	½	6	81.14	½	7	01.81	½	7	22.48

BULLION ROOMS, ASSAY OFFICE, CHEMICAL LABORATORY AND ORE FLOORS,
524 Sacramento Street, San Francisco.

TABLE OF THE VALUE OF
GOLD
Per OUNCE TROY, AT DIFFERENT DEGREES OF FINENESS

-BY-

THOMAS PRICE.

FINE	DOLLARS	CENTS	FINE	DOLLARS	CENTS	FINE	DOLLARS	CENTS	FINE	DOLLARS	CENTS	FINE	DOLLARS	CENTS
350	7	23.51	360	7	44.19	370	7	64.86	380	7	85.53	390	8	06.20
½	7	24.55	½	7	45.22	½	7	65.89	½	7	86.56	½	8	07.24
351	7	25.58	361	7	46.25	371	7	66.93	381	7	87.60	391	8	08.27
½	7	26.61	½	7	47.29	½	7	67.96	½	7	88.63	½	8	09.30
352	7	27.65	362	7	48.32	372	7	68.99	382	7	89.66	392	8	10.34
½	7	28.68	½	7	49.35	½	7	70.03	½	7	90.70	½	8	11.37
353	7	29.72	363	7	50.39	373	7	71.06	383	7	91.73	393	8	12.40
½	7	30.75	½	7	51.42	½	7	72.09	½	7	92.76	½	8	13.44
354	7	31.78	364	7	52.45	374	7	73.13	384	7	93.80	394	8	14.47
½	7	32.82	½	7	53.49	½	7	74.16	½	7	94.83	½	8	15.50
355	7	33.85	365	7	54.52	375	7	75.19	385	7	95.87	395	8	16.54
½	7	34.88	½	7	55.56	½	7	76.23	½	7	96.90	½	8	17.57
356	7	35.92	366	7	56.59	376	7	77.26	386	7	97.93	396	8	18.60
½	7	36.95	½	7	57.62	½	7	78.29	½	7	98.97	½	8	19.64
357	7	37.98	367	7	58.66	377	7	79.32	387	8	00.00	397	8	20.67
½	7	39.02	½	7	59.69	½	7	80.36	½	8	01.03	½	8	21.71
358	7	40.05	368	7	60.72	378	7	81.39	388	8	02.07	398	8	22.74
½	7	41.09	½	7	61.76	½	7	82.43	½	8	03.10	½	8	23.77
359	7	42.12	369	7	62.79	379	7	83.46	389	8	04.13	399	8	24.81
½	7	43.15	½	7	63.82	½	7	84.50	½	8	05.17	½	8	25.84

BULLION ROOMS, ASSAY OFFICE, CHEMICAL LABORATORY AND ORE FLOORS,

524 Sacramento Street, San Francisco.

TABLE OF THE VALUE OF
GOLD

Per OUNCE TROY, AT DIFFERENT DEGREES OF FINENESS

–BY–

THOMAS PRICE.

FINE.	DOLLARS.	CENTS.	FINE.	DOLLARS.	CENTS.	FINE.	DOLLARS.	CENTS.	FINE.	DOLLARS.	CENTS.	FINE.	DOLLARS.	CENTS.
400	8	26.87	**410**	8	47.55	**420**	8	68.22	**430**	8	88.89	**440**	9	09.56
½	8	27.91	½	8	48.58	½	8	69.25	½	8	89.92	½	9	10.59
401	8	28.94	411	8	49.61	421	8	70.28	431	8	90.96	441	9	11.63
½	8	29.97	½	8	50.65	½	8	71.32	½	8	91.99	½	9	12.66
402	8	31.01	412	8	51.68	422	8	72.35	432	8	93.02	442	9	13.70
½	8	32.04	½	8	52.71	½	8	73.39	½	8	94.06	½	9	14.73
403	8	33.07	413	8	53.75	423	8	74.42	433	8	95.09	443	9	15.76
½	8	34.11	½	8	54.78	½	8	75.45	½	8	96.12	½	9	16.80
404	8	35.14	414	8	55.81	424	8	76.49	434	8	97.16	444	9	17.83
½	8	36.18	½	8	56.85	½	8	77.52	½	8	98.19	½	9	18.86
405	8	37.21	415	8	57.88	425	8	78.55	435	8	99.22	445	9	19.90
½	8	38.24	½	8	58.91	½	8	79.59	½	9	00.26	½	9	20.93
406	8	39.28	416	8	59.95	426	8	80.62	436	9	01.29	446	9	21.96
½	8	40.31	½	8	60.98	½	8	81.65	½	9	02.33	½	9	23.00
407	8	41.34	417	8	62.02	427	8	82.69	437	9	03.36	447	9	24.03
½	8	42.38	½	8	63.05	½	8	83.72	½	9	04.39	½	9	25.06
408	8	43.41	418	8	64.08	428	8	84.75	438	9	05.43	448	9	26.10
½	8	44.44	½	8	65.12	½	8	85.79	½	9	06.46	½	9	27.13
409	8	45.48	419	8	66.15	429	8	86.82	439	9	07.49	449	9	28.17
½	8	46.51	½	8	67.18	½	8	87.86	½	9	08.53	½	9	29.20

BULLION ROOMS, ASSAY OFFICE, CHEMICAL LABORATORY AND ORE FLOORS,

524 Sacramento Street, San Francisco.

TABLE OF THE VALUE OF
GOLD
Per OUNCE TROY, AT DIFFERENT DEGREES OF FINENESS
-BY-
THOMAS PRICE.

FINE	DOLLARS	CENTS	FINE	DOLLARS	CENTS	FINE	DOLLARS	CENTS	FINE	DOLLARS	CENTS	FINE	DOLLARS	CENTS
450	9	30.23	460	9	50.90	470	9	71.58	480	9	92.25	490	10	12.92
½	9	31.27	½	9	51.94	½	9	72.61	½	9	93.28	½	10	13.95
451	9	32.30	461	9	52.97	471	9	73.64	481	9	94.32	491	10	14.99
½	9	33.33	½	9	54.01	½	9	74.68	½	9	95.35	½	10	16.02
452	9	34.37	462	9	55.04	472	9	75.71	482	9	96.38	492	10	17.05
½	9	35.40	½	9	56.07	½	9	76.74	½	9	97.42	½	10	18.09
453	9	36.43	463	9	57.11	473	9	77.78	483	9	98.45	493	10	19.12
½	9	37.47	½	9	58.14	½	9	78.81	½	9	99.48	½	10	20.16
454	9	38.50	464	9	59.17	474	9	79.84	484	10	00.52	494	10	21.19
½	9	39.53	½	9	60.21	½	9	80.88	½	10	01.55	½	10	22.22
455	9	40.57	465	9	61.24	475	9	81.91	485	10	02.58	495	10	23.26
½	9	41.60	½	9	62.27	½	9	82.95	½	10	03.62	½	10	24.29
456	9	42.64	466	9	63.31	476	9	83.98	486	10	04.65	496	10	25.32
½	9	43.67	½	9	64.34	½	9	85.01	½	10	05.68	½	10	26.36
457	9	44.70	467	9	65.37	477	9	86.05	487	10	06.72	497	10	27.39
½	9	45.74	½	9	66.41	½	9	87.08	½	10	07.75	½	10	28.42
458	9	46.77	468	9	67.44	478	9	88.11	488	10	08.79	498	10	29.46
½	9	47.80	½	9	68.48	½	9	89.15	½	10	09.82	½	10	30.49
459	9	48.84	469	9	69.51	479	9	90.18	489	10	10.85	499	10	31.52
½	9	49.87	½	9	70.54	½	9	91.21	½	10	11.89	½	10	32.56

BULLION ROOMS, ASSAY OFFICE, CHEMICAL LABORATORY AND ORE FLOORS,
524 Sacramento Street, San Francisco.

TABLE OF THE VALUE OF
GOLD
Per OUNCE TROY, AT DIFFERENT DEGREES OF FINENESS
—BY—
THOMAS PRICE.

FINE.	DOLLARS.	CENTS.	FINE.	DOLLARS.	CENTS.	FINE.	DOLLARS.	CENTS.	FINE.	DOLLARS.	CENTS.	FINE.	DOLLARS.	CENTS.
500	10	33.59	510	10	54.26	520	10	74.94	530	10	95.61	540	11	16.28
½	10	34.63	½	10	55.30	½	10	75.97	½	10	96.64	½	11	17.31
501	10	35.66	511	10	56.33	521	10	77.00	531	10	97.67	541	11	18.35
½	10	36.69	½	10	57.36	½	10	78.04	½	10	98.71	½	11	19.38
502	10	37.73	512	10	58.40	522	10	79.07	532	10	99.74	542	11	20.41
½	10	38.76	½	10	59.43	½	10	80.10	½	11	00.78	½	11	21.45
503	10	39.79	513	10	60.47	523	10	81.14	533	11	01.81	543	11	22.48
½	10	40.83	½	10	61.50	½	10	82.17	½	11	02.84	½	11	23.51
504	10	41.86	514	10	62.53	524	10	83.20	534	11	03.88	544	11	24.55
½	10	42.89	½	10	63.57	½	10	84.24	½	11	04.91	½	11	25.58
505	10	43.93	515	10	64.60	525	10	85.27	535	11	05.94	545	11	26.61
½	10	44.96	½	10	65.63	½	10	86.30	½	11	06.98	½	11	27.65
506	10	45.99	516	10	66.67	526	10	87.34	536	11	08.01	546	11	28.68
½	10	47.03	½	10	67.70	½	10	88.37	½	11	09.04	½	11	29.72
507	10	48.06	517	10	68.73	527	10	89.41	537	11	10.08	547	11	30.75
½	10	49.10	½	10	69.77	½	10	90.44	½	11	11.11	½	11	31.78
508	10	50.13	518	10	70.80	528	10	91.47	538	11	12.14	548	11	32.82
½	10	51.16	½	10	71.83	½	10	92.51	½	11	13.18	½	11	33.85
509	10	52.20	519	10	72.87	529	10	93.54	539	11	14.21	549	11	34.88
½	10	53.23	½	10	73.90	½	10	94.57	½	11	15.25	½	11	35.92

BULLION ROOMS, ASSAY OFFICE, CHEMICAL LABORATORY AND ORE FLOORS,

524 Sacramento Street, San Francisco.

TABLE OF THE VALUE OF
GOLD
Per OUNCE TROY, AT DIFFERENT DEGREES OF FINENESS

-BY-

THOMAS PRICE.

FINE.	DOLLARS.	CENTS.	FINE.	DOLLARS.	CENTS.	FINE.	DOLLARS.	CENTS.	FINE.	DOLLARS.	CENTS.	FINE.	DOLLARS.	CENTS.
550	11	36.95	560	11	57.62	570	11	78.29	580	11	98.97	590	12	19.64
½	11	37.98	½	11	58.66	½	11	79.33	½	12	00.00	½	12	20.67
551	11	39.02	561	11	59.69	571	11	80.36	581	12	01.03	591	12	21.71
½	11	40.05	½	11	60.72	½	11	81.40	½	12	02.07	½	12	22.74
552	11	41.09	562	11	61.76	572	11	82.43	582	12	03.10	592	12	23.77
½	11	42.12	½	11	62.79	½	11	83.46	½	12	04.13	½	12	24.81
553	11	43.15	563	11	63.82	573	11	84.50	583	12	05.17	593	12	25.84
½	11	44.19	½	11	64.86	½	11	85.53	½	12	06.20	½	12	26.87
554	11	45.22	564	11	65.89	574	11	86.56	584	12	07.24	594	12	27.91
½	11	46.25	½	11	66.93	½	11	87.60	½	12	08.27	½	12	28.94
555	11	47.29	565	11	67.96	575	11	88.63	585	12	09.30	595	12	29.97
½	11	48.32	½	11	68.99	½	11	89.66	½	12	10.34	½	12	31.01
556	11	49.35	566	11	70.03	576	11	90.70	586	12	11.37	596	12	32.04
½	11	50.39	½	11	71.06	½	11	91.73	½	12	12.40	½	12	33.07
557	11	51.42	567	11	72.09	577	11	92.76	587	12	13.44	597	12	34.11
½	11	52.45	½	11	73.13	½	11	93.80	½	12	14.47	½	12	35.14
558	11	53.49	568	11	74.16	578	11	94.83	588	12	15.50	598	12	36.18
½	11	54.52	½	11	75.19	½	11	95.87	½	12	16.54	½	12	37.21
559	11	55.56	569	11	76.23	579	11	96.90	589	12	17.57	599	12	38.24
½	11	56.59	½	11	77.26	½	11	97.93	½	12	18.60	½	12	39.28

BULLION ROOMS, ASSAY OFFICE, CHEMICAL LABORATORY AND ORE FLOORS,

524 Sacramento Street, San Francisco.

TABLE OF THE VALUE OF
GOLD
Per OUNCE TROY, AT DIFFERENT DEGREES OF FINENESS
-BY-
THOMAS PRICE.

FINE.	DOLLARS.	CENTS.	FINE.	DOLLARS.	CENTS.	FINE.	DOLLARS.	CENTS.	FINE.	DOLLARS.	CENTS.	FINE.	DOLLARS.	CENTS.
600	12	40.31	610	12	60.98	620	12	81.65	630	13	02.33	640	13	23.00
½	12	41.34	½	12	62.02	½	12	82.69	½	13	03.36	½	13	24.03
601	12	42.38	611	12	63.05	621	12	83.72	631	13	04.39	641	13	25.06
½	12	43.41	½	12	64.08	½	12	84.75	½	13	05.43	½	13	26.10
602	12	44.44	612	12	65.12	622	12	85.79	632	13	06.46	642	13	27.13
½	12	45.48	½	12	66.15	½	12	86.82	½	13	07.49	½	13	28.17
603	12	46.51	613	12	67.18	623	12	87.86	633	13	08.53	643	13	29.20
½	12	47.55	½	12	68.22	½	12	88.89	½	13	09.56	½	13	30.23
604	12	48.58	614	12	69.25	624	12	89.92	634	13	10.59	644	13	31.27
½	12	49.61	½	12	70.28	½	12	90.96	½	13	11.63	½	13	32.30
605	12	50.65	615	12	71.32	625	12	91.99	635	13	12.66	645	13	33.33
½	12	51.68	½	12	72.35	½	12	93.02	½	13	13.70	½	13	34.37
606	12	52.71	616	12	73.39	626	12	94.06	636	13	14.73	646	13	35.40
½	12	53.75	½	12	74.42	½	12	95.09	½	13	15.76	½	13	36.43
607	12	54.78	617	12	75.45	627	12	96.12	637	13	16.80	647	13	37.47
½	12	55.81	½	12	76.49	½	12	97.16	½	13	17.83	½	13	38.50
608	12	56.85	618	12	77.52	628	12	98.19	638	13	18.86	648	13	39.53
½	12	57.88	½	12	78.55	½	12	99.22	½	13	19.90	½	13	40.57
609	12	58.91	619	12	79.59	629	13	00.26	639	13	20.93	649	13	41.60
½	12	59.95	½	12	80.62	½	13	01.29	½	13	21.96	½	13	42.64

BULLION ROOMS, ASSAY OFFICE, CHEMICAL LABORATORY AND ORE FLOORS,
524 Sacramento Street, San Francisco.

TABLE OF THE VALUE OF
GOLD
Per OUNCE TROY, AT DIFFERENT DEGREES OF FINENESS
-BY-
THOMAS PRICE.

FINE.	DOLLARS.	CENTS.	FINE.	DOLLARS.	CENTS.	FINE.	DOLLARS.	CENTS.	FINE.	DOLLARS.	CENTS.	FINE.	DOLLARS.	CENTS.
650	13	43.67	660	13	64.34	670	13	85.01	680	14	05.68	690	14	26.36
½	13	44.70	½	13	65.37	½	13	86.05	½	14	06.72	½	14	27.39
651	13	45.74	661	13	66.41	671	13	87.08	681	14	07.75	691	14	28.42
½	13	46.77	½	13	67.44	½	13	88.11	½	14	08.79	½	14	29.46
652	13	47.80	662	13	68.48	672	13	89.15	682	14	09.82	692	14	30.49
½	13	48.84	½	13	69.51	½	13	90.18	½	14	10.85	½	14	31.52
653	13	49.87	663	13	70.54	673	13	91.21	683	14	11.89	693	14	32.56
½	13	50.90	½	13	71.58	½	13	92.25	½	14	12.92	½	14	33.59
654	13	51.94	664	13	72.61	674	13	93.28	684	14	13.95	694	14	34.63
½	13	52.97	½	13	73.64	½	13	94.32	½	14	14.99	½	14	35.66
655	13	54.01	665	13	74.68	675	13	95.35	685	14	16.02	695	14	36.69
½	13	55.04	½	13	75.71	½	13	96.38	½	14	17.05	½	14	37.73
656	13	56.07	666	13	76.74	676	13	97.42	686	14	18.09	696	14	38.76
½	13	57.11	½	13	77.78	½	13	98.45	½	14	19.12	½	14	39.79
657	13	58.14	667	13	78.81	677	13	99.48	687	14	20.16	697	14	40.83
½	13	59.17	½	13	79.84	½	14	00.52	½	14	21.19	½	14	41.86
658	13	60.21	668	13	80.88	678	14	01.55	688	14	22.22	698	14	42.89
½	13	61.24	½	13	81.91	½	14	02.58	½	14	23.26	½	14	43.93
659	13	62.27	669	13	82.95	679	14	03.62	689	14	24.29	699	14	44.96
½	13	63.31	½	13	83.98	½	14	04.65	½	14	25.32	½	14	45.99

BULLION ROOMS, ASSAY OFFICE, CHEMICAL LABORATORY AND ORE FLOORS,
524 Sacramento Street, San Francisco.

TABLE OF THE VALUE OF

GOLD

Per OUNCE TROY, AT DIFFERENT DEGREES OF FINENESS

-BY-

THOMAS PRICE.

FINE.	DOLLARS.	CENTS.	FINE.	DOLLARS.	CENTS.	FINE.	DOLLARS.	CENTS.	FINE.	DOLLARS.	CENTS.	FINE.	DOLLARS.	CENTS.
700	14	47.03	**710**	14	67.70	**720**	14	88.37	**730**	15	09.04	**740**	15	29.72
½	14	48.06	½	14	68.73	½	14	89.41	½	15	10.08	½	15	30.75
701	14	49.10	711	14	69.76	721	14	90.44	731	15	11.11	741	15	31.78
½	14	50.13	½	14	70.80	½	14	91.47	½	15	12.14	½	15	32.82
702	14	51.16	712	14	71.83	722	14	92.51	732	15	13.18	742	15	33.85
½	14	52.20	½	14	72.87	½	14	93.54	½	15	14.21	½	15	34.88
703	14	53.23	713	14	73.90	723	14	94.57	733	15	15.25	743	15	35.92
½	14	54.26	½	14	74.94	½	14	95.61	½	15	16.28	½	15	36.95
704	14	55.30	714	14	75.97	724	14	96.64	734	15	17.31	744	15	37.98
½	14	56.33	½	14	77.00	½	14	97.67	½	15	18.35	½	15	39.02
705	14	57.36	715	14	78.04	725	14	98.71	735	15	19.38	745	15	40.05
½	14	58.40	½	14	79.07	½	14	99.74	½	15	20.41	½	15	41.09
706	14	59.43	716	14	80.10	726	15	00.78	736	15	21.45	746	15	42.12
½	14	60.47	½	14	81.14	½	15	01.81	½	15	22.48	½	15	43.15
707	14	61.50	717	14	82.17	727	15	02.84	737	15	23.51	747	15	44.18
½	14	62.53	½	14	83.20	½	15	03.88	½	15	24.55	½	15	45.22
708	14	63.57	718	14	84.24	728	15	04.91	738	15	25.58	748	15	46.25
½	14	64.60	½	14	85.27	½	15	05.94	½	15	26.61	½	15	47.29
709	14	65.63	719	14	86.30	729	15	06.98	739	15	27.65	749	15	48.32
½	14	66.67	½	14	87.34	½	15	08.01	½	15	28.68	½	15	49.35

BULLION ROOMS, ASSAY OFFICE, CHEMICAL LABORATORY AND ORE FLOORS,

524 Sacramento Street, San Francisco.

TABLE OF THE VALUE OF
GOLD
Per OUNCE TROY, AT DIFFERENT DEGREES OF FINENESS
-BY-
THOMAS PRICE.

FINE.	DOLLARS.	CENTS.	FINE.	DOLLARS.	CENTS.	FINE.	DOLLARS.	CENTS.	FINE.	DOLLARS.	CENTS.	FINE.	DOLLARS.	CENTS.
750	15	50.39	760	15	71.06	770	15	91.73	780	16	12.40	790	16	33.07
½	15	51.42	½	15	72.09	½	15	92.76	½	16	13.44	½	16	34.11
751	15	52.45	761	15	73.13	771	15	93.80	781	16	14.47	791	16	35.14
½	15	53.49	½	15	74.16	½	15	94.83	½	16	15.50	½	16	36.18
752	15	54.52	762	15	75.19	772	15	95.87	782	16	16.54	792	16	37.21
½	15	55.56	½	15	76.23	½	15	96.90	½	16	17.57	½	16	38.24
753	15	56.59	763	15	77.26	773	15	97.93	783	16	18.60	793	16	39.28
½	15	57.62	½	15	78.29	½	15	98.97	½	16	19.64	½	16	40.31
754	15	58.66	764	15	79.33	774	16	00.00	784	16	20.67	794	16	41.34
½	15	59.69	½	15	80.36	½	16	01.03	½	16	21.71	½	16	42.38
755	15	60.72	765	15	81.40	775	16	02.07	785	16	22.74	795	16	43.41
½	15	61.76	½	15	82.43	½	16	03.10	½	16	23.77	½	16	44.44
756	15	62.79	766	15	83.46	776	16	04.13	786	16	24.81	796	16	45.48
½	15	63.82	½	15	84.50	½	16	05.17	½	16	25.84	½	16	46.51
757	15	64.86	767	15	85.53	777	16	06.20	787	16	26.87	797	16	47.55
½	15	65.89	½	15	86.56	½	16	07.24	½	16	27.91	½	16	48.58
758	15	66.93	768	15	87.60	778	16	08.27	788	16	28.94	798	16	49.61
½	15	67.96	½	15	88.63	½	16	09.30	½	16	29.97	½	16	50.65
759	15	68.99	769	15	89.66	779	16	10.34	789	16	31.01	799	16	51.68
½	15	70.03	½	15	90.70	½	16	11.37	½	16	32.04	½	16	52.71

BULLION ROOMS, ASSAY OFFICE, CHEMICAL LABORATORY AND ORE FLOORS,
524 Sacramento Street, San Francisco.

TABLE OF THE VALUE OF
GOLD
Per OUNCE TROY, AT DIFFERENT DEGREES OF FINENESS

–BY–

THOMAS PRICE.

FINE.	DOLLARS.	CENTS.	FINE.	DOLLARS.	CENTS.	FINE.	DOLLARS.	CENTS.	FINE.	DOLLARS.	CENTS.	FINE.	DOLLARS.	CENTS.
800	16	53.75	**810**	16	74.42	**820**	16	95.09	**830**	17	15.76	**840**	17	36.43
½	16	54.78	½	16	75.45	½	16	96.12	½	17	16.80	½	17	37.47
801	16	55.81	811	16	76.49	821	16	97.16	831	17	17.83	841	17	38.50
½	16	56.85	½	16	77.52	½	16	98.19	½	17	18.86	½	17	39.53
802	16	57.88	812	16	78.55	822	16	99.22	832	17	19.90	842	17	40.57
½	16	58.91	½	16	79.59	½	17	00.26	½	17	20.93	½	17	41.60
803	16	59.95	813	16	80.62	823	17	01.29	833	17	21.96	843	17	42.64
½	16	60.98	½	16	81.65	½	17	02.33	½	17	23.00	½	17	43.67
804	16	62.02	814	16	82.69	824	17	03.36	834	17	24.03	844	17	44.70
½	16	63.05	½	16	83.72	½	17	04.39	½	17	25.06	½	17	45.74
805	16	64.08	815	16	84.75	825	17	05.43	835	17	26.10	845	17	46.77
½	16	65.12	½	16	85.79	½	17	06.46	½	17	27.13	½	17	47.80
806	16	66.15	816	16	86.82	826	17	07.49	836	17	28.17	846	17	48.84
½	16	67.18	½	16	87.86	½	17	08.53	½	17	29.20	½	17	49.87
807	16	68.22	817	16	88.89	827	17	09.56	837	17	30.23	847	17	50.90
½	16	69.25	½	16	89.92	½	17	10.59	½	17	31.27	½	17	51.94
808	16	70.28	818	16	90.96	828	17	11.63	838	17	32.30	848	17	52.97
½	16	71.32	½	16	91.99	½	17	12.66	½	17	33.33	½	17	54.01
809	16	72.35	819	16	93.02	829	17	13.70	839	17	34.37	849	17	55.04
½	16	73.39	½	16	94.06	½	17	14.73	½	17	35.40	½	17	56.07

BULLION ROOMS, ASSAY OFFICE, CHEMICAL LABORATORY AND ORE FLOORS,
524 Sacramento Street, San Francisco.

TABLE OF THE VALUE OF
GOLD
Per OUNCE TROY, AT DIFFERENT DEGREES OF FINENESS
–BY–
THOMAS PRICE.

FINE.	DOLLARS.	CENTS.	FINE.	DOLLARS.	CENTS.	FINE.	DOLLARS.	CENTS.	FINE.	DOLLARS.	CENTS.	FINE.	DOLLARS.	CENTS.
850	17	57.11	860	17	77.78	870	17	98.45	880	18	19.12	890	18	39.79
½	17	58.14	½	17	78.81	½	17	99.48	½	18	20.16	½	18	40.83
851	17	59.17	861	17	79.84	871	18	00.52	881	18	21.19	891	18	41.86
½	17	60.21	½	17	80.88	½	18	01.55	½	18	22.22	½	18	42.89
852	17	61.24	862	17	81.91	872	18	02.58	882	18	23.26	892	18	43.93
½	17	62.27	½	17	82.95	½	18	03.62	½	18	24.29	½	18	44.96
853	17	63.31	863	17	83.98	873	18	04.65	883	18	25.32	893	18	45.99
½	17	64.34	½	17	85.01	½	18	05.68	½	18	26.36	½	18	47.03
854	17	65.37	864	17	86.05	874	18	06.72	884	18	27.39	894	18	48.06
½	17	66.41	½	17	87.08	½	18	07.75	½	18	28.42	½	18	49.10
855	17	67.44	865	17	88.11	875	18	08.79	885	18	29.46	895	18	50.13
½	17	68.48	½	17	89.15	½	18	09.82	½	18	30.49	½	18	51.16
856	17	69.51	866	17	90.18	876	18	10.85	886	18	31.52	896	18	52.20
½	17	70.54	½	17	91.21	½	18	11.89	½	18	32.56	½	18	53.23
857	17	71.58	867	17	92.25	877	18	12.92	887	18	33.59	897	18	54.26
½	17	72.61	½	17	93.28	½	18	13.95	½	18	34.63	½	18	55.30
858	17	73.64	868	17	94.32	878	18	14.99	888	18	35.66	898	18	56.33
½	17	74.68	½	17	95.35	½	18	16.02	½	18	36.69	½	18	57.36
859	17	75.71	869	17	96.38	879	18	17.05	889	18	37.73	899	18	58.40
½	17	76.74	½	17	97.42	½	18	18.09	½	18	38.76	½	18	59.43

BULLION ROOMS, ASSAY OFFICE, CHEMICAL LABORATORY AND ORE FLOORS,
524 Sacramento Street, San Francisco.

TABLE OF THE VALUE OF
GOLD

Per OUNCE TROY, AT DIFFERENT DEGREES OF FINENESS

–BY–

THOMAS PRICE.

FINE.	DOLLARS.	CENTS.	FINE.	DOLLARS.	CENTS.	FINE.	DOLLARS.	CENTS.	FINE.	DOLLARS.	CENTS.	FINE.	DOLLARS.	CENTS.
900	18	60.46	**910**	18	81.14	**920**	19	01.81	**930**	19	22.48	**940**	19	43.15
½	18	61.50	½	18	82.17	½	19	02.84	½	19	23.51	½	19	44.19
901	18	62.53	911	18	83.20	921	19	03.88	931	19	24.55	941	19	45.22
½	18	63.57	½	18	84.24	½	19	04.91	½	19	25.58	½	19	46.25
902	18	64.60	912	18	85.27	922	19	05.94	932	19	26.61	942	19	47.29
½	18	65.63	½	18	86.30	½	19	06.98	½	19	27.65	½	19	48.32
903	18	66.67	913	18	87.34	923	19	08.01	933	19	28.68	943	19	49.35
½	18	67.70	½	18	88.37	½	19	09.04	½	19	29.72	½	19	50.39
904	18	68.73	914	18	89.41	924	19	10.08	934	19	30.75	944	19	51.42
½	18	69.77	½	18	90.44	½	19	11.11	½	19	31.78	½	19	52.45
905	18	70.80	915	18	91.47	925	19	12.14	935	19	32.82	945	19	53.49
½	18	71.83	½	18	92.51	½	19	13.18	½	19	33.85	½	19	54.52
906	18	72.87	916	18	93.54	926	19	14.21	936	19	34.88	946	19	55.56
½	18	73.90	½	18	94.57	½	19	15.25	½	19	35.92	½	19	56.59
907	18	74.94	917	18	95.61	927	19	16.28	937	19	36.95	947	19	57.62
½	18	75.97	½	18	96.64	½	19	17.31	½	19	37.98	½	19	58.66
908	18	77.00	918	18	97.67	928	19	18.35	938	19	39.02	948	19	59.69
½	18	78.04	½	18	98.71	½	19	19.38	½	19	40.05	½	19	60.72
909	18	79.07	919	18	99.74	929	19	20.41	939	19	41.08	949	19	61.76
½	18	80.10	½	19	00.78	½	19	21.45	½	19	42.12	½	19	62.79

BULLION ROOMS, ASSAY OFFICE, CHEMICAL LABORATORY AND ORE FLOORS,

524 Sacramento Street, San Francisco.

TABLE OF THE VALUE OF
GOLD
Per OUNCE TROY, AT DIFFERENT DEGREES OF FINENESS

–BY–

THOMAS PRICE.

FINE.	DOLLARS.	CENTS.	FINE.	DOLLARS.	CENTS.	FINE.	DOLLARS.	CENTS.	FINE.	DOLLARS.	CENTS.	FINE.	DOLLARS.	CENTS.
950	19	63.82	960	19	84.50	970	20	05.17	980	20	25.84	990	20	46.51
½	19	64.86	½	19	85.53	½	20	06.20	½	20	26.87	½	20	47.55
951	19	65.89	961	19	86.56	971	20	07.23	981	20	27.91	991	20	48.58
½	19	66.93	½	19	87.60	½	20	08.27	½	20	28.94	½	20	49.61
952	19	67.96	962	19	88.63	972	20	09.30	982	20	29.97	992	20	50.65
½	19	68.99	½	19	89.66	½	20	10.34	½	20	31.01	½	20	51.68
953	19	70.03	963	19	90.70	973	20	11.37	983	20	32.04	993	20	52.71
½	19	71.06	½	19	91.73	½	20	12.40	½	20	33.07	½	20	53.75
954	19	72.09	964	19	92.76	974	20	13.44	984	20	34.11	994	20	54.78
½	19	73.13	½	19	93.80	½	20	14.47	½	20	35.14	½	20	55.81
955	19	74.16	965	19	94.83	975	20	15.50	985	20	36.18	995	20	56.85
½	19	75.19	½	19	95.87	½	20	16.54	½	20	37.21	½	20	57.88
956	19	76.23	966	19	96.90	976	20	17.57	986	20	38.24	996	20	58.91
½	19	77.26	½	19	97.93	½	20	18.60	½	20	39.28	½	20	59.95
957	19	78.29	967	19	98.97	977	20	19.64	987	20	40.31	997	20	60.98
½	19	79.33	½	20	00.00	½	20	20.67	½	20	41.34	½	20	62.02
958	19	80.36	968	20	01.03	978	20	21.70	988	20	42.38	998	20	63.05
½	19	81.40	½	20	02.07	½	20	22.74	½	20	43.41	½	20	64.08
959	19	82.43	969	20	03.10	979	20	23.77	989	20	44.44	999	20	65.12
½	19	83.46	½	20	04.13	½	20	24.81	½	20	45.48	½	20	66.15

1000	20	67.18

BULLION ROOMS, ASSAY OFFICE, CHEMICAL LABORATORY AND ORE FLOORS,
524 Sacramento Street, San Francisco.

EXPLANATION OF SILVER TABLE.

The values of silver per oz. in the following tables are computed from the formula that 99 ozs. of pure silver (1000 fine) are worth $128.00. Hence, 1 oz. is worth $1.29.29, etc., and the ₁∕₁₀₀₀ (.001) of an oz. is worth $.000.129.29. And that 11 ozs. of standard silver (900 fine) are worth $12.80, and hence 1 oz. standard silver is worth $1.16.36. These values (i. e. $1.29 for fine silver and $1.16 for standard silver) are the *intrinsic* values of silver, being the values at which silver is equal to gold, dollar for dollar, or as $1 is to 15.98837, etc. Silver, however, is usually at a premium or discount, which varies with the supply and demand.

If one oz. of pure silver (1000 fine) is worth $1.29.29, 1 oz. of silver 900 fine is worth $1.16.36 (viz., $1.29.29×900.) Hence, a silver bar weighing 1000 ozs. and containing 900 parts silver, or 900 fine, multiplied by $1.16.36, equals $1,163.60.

EXAMPLE illustrating the application of both the Gold and Silver Tables, viz: Take a silver bar of say 100 ozs.—900 fine of silver and 90 fine of gold: by reference to the Silver Table, opposite 900, the value of 1 oz. at this fineness is $1.16.36, and for 100 ozs. the value therefore is..$116 36

And opposite 90 in the Gold Table, the value of 1 oz. is $1.86.05, and for 100 ozs..... 186 05

The total value of the bar is........................$302 41

Gold bars are usually stamped with the fineness and value only of the gold contained and by a rule established by our bankers at an early date as a more convenient basis for their general transactions, it is customary to allow for ten parts base metal; hence, in a bar of 100 ozs. gold stamped 900 fine, value $1,860.46, it is understood that the fineness of the silver is .90, and the price to be paid for the bar is governed by the proportion of silver contained. [See Tables following for examples.]

Calculations of the value of metal may also be ascertained by reducing the proportions to fine gold and fine silver, and multiplying by the value per oz. of pure gold and pure silver. The following rule is applicable, viz: Gross weight multiplied by fineness, divided by 1000, gives net weight of pure metal.

EXAMPLE.—A bar 500 ozs. gross, 820 fine of GOLD, 170 fine of SILVER.

500×820=410 ozs. pure gold, at $20.67.18....................$8,475 44
500×170 = 85 ozs. pure silver, at $1.29.29..................... 109 89

Total value.......................................$8,585 33

TABLE OF THE VALUE OF
SILVER

Per OUNCE TROY, AT DIFFERENT DEGREES OF FINENESS

–BY–

THOMAS PRICE.

FINE.	DOLLARS.	CENTS.	FINE.	DOLLARS.	CENTS.	FINE.	DOLLARS.	CENTS.	FINE.	DOLLARS.	CENTS.	FINE.	DOLLARS.	CENTS.
00	0	00.00	20		02.59	40		05.17	60		07.76	80		10.34
	1	00.13		21	02.72		41	05.30		61	07.89		81	10.47
	2	00.26		22	02.84		42	05.43		62	08.02		82	10.60
	3	00.39		23	02.97		43	05.56		63	08.15		83	10.73
	4	00.52		24	03.10		44	05.69		64	08.27		84	10.86
	5	00.65		25	03.23		45	05.82		65	08.40		85	10.99
	6	00.78		26	03.36		46	05.95		66	08.53		86	11.12
	7	00.90		27	03.49		47	06.08		67	08.66		87	11.25
	8	01.03		28	03.62		48	06.21		68	08.79		88	11.38
	9	01.16		29	03.75		49	06.34		69	08.92		89	11.51
10		01.29	30		03.88	50		06.46	70		09.05	90		11.64
	11	01.42		31	04.01		51	06.59		71	09.18		91	11.77
	12	01.55		32	04.14		52	06.72		72	09.31		92	11.89
	13	01.68		33	04.27		53	06.85		73	09.44		93	12.02
	14	01.81		34	04.40		54	06.98		74	09.57		94	12.15
	15	01.94		35	04.53		55	07.11		75	09.70		95	12.28
	16	02.07		36	04.65		56	07.24		76	09.83		96	12.41
	17	02.20		37	04.78		57	07.37		77	09.96		97	12.54
	18	02.33		38	04.91		58	07.50		78	10.08		98	12.67
	19	02.46		39	05.04		59	07.63		79	10.21		99	12.80

BULLION ROOMS, ASSAY OFFICE, CHEMICAL LABORATORY AND ORE FLOORS,

524 Sacramento Street, San Francisco.

TABLE OF THE VALUE OF
SILVER
Per OUNCE TROY, AT DIFFERENT DEGREES OF FINENESS

–BY–

THOMAS PRICE.

FINE.	DOLLARS. CENTS.	FINE.	DOLLARS. CENTS.	FINE.	DOLLARS. CENTS.	FINE.	DOLLARS. CENTS.	FINE.	DOLLARS. CENTS.
100	12.93	120	15.52	140	18.10	160	20.69	180	23.27
101	13.06	121	15.64	141	18.23	161	20.82	181	23.40
102	13.19	122	15.77	142	18.36	162	20.95	182	23.53
103	13.32	123	15.90	143	18.49	163	21.07	183	23.66
104	13.45	124	16.03	144	18.62	164	21.20	184	23.79
105	13.58	125	16.16	145	18.75	165	21.33	185	23.92
106	13.70	126	16.29	146	18.88	166	21.46	186	24.05
107	13.83	127	16.42	147	19.01	167	21.59	187	24.18
108	13.96	128	16.55	148	19.14	168	21.72	188	24.31
109	14.09	129	16.68	149	19.26	169	21.85	189	24.44
110	14.22	130	16.81	150	19.39	170	21.98	190	24.57
111	14.35	131	16.94	151	19.52	171	22.11	191	24.69
112	14.48	132	17.07	152	19.65	172	22.24	192	24.82
113	14.61	133	17.20	153	19.78	173	22.37	193	24.95
114	14.74	134	17.33	154	19.91	174	22.50	194	25.08
115	14.87	135	17.45	155	20.04	175	22.63	195	25.21
116	15.00	136	17.58	156	20.17	176	22.76	196	25.34
117	15.13	137	17.71	157	20.30	177	22.88	197	25.47
118	15.26	138	17.84	158	20.43	178	23.01	198	25.60
119	15.39	139	17.97	159	20.56	179	23.14	199	25.73

Bullion Rooms, Assay Office, Chemical Laboratory and Ore Floors,
524 Sacramento Street, San Francisco.

TABLE OF THE VALUE OF
SILVER
Per OUNCE TROY, AT DIFFERENT DEGREES OF FINENESS
–BY–
THOMAS PRICE.

FINE.	DOLLARS.	CENTS.	FINE.	DOLLARS.	CENTS.	FINE.	DOLLARS.	CENTS.	FINE.	DOLLARS.	CENTS.	FINE.	DOLLARS.	CENTS.
200		25.86	**220**		28.44	**240**		31.03	**260**		33.62	**280**		36.20
201		25.99	221		28.57	241		31.16	261		33.75	281		36.33
202		26.12	222		28.70	242		31.29	262		33.87	282		36.46
203		26.25	223		28.83	243		31.42	263		34.00	283		36.59
204		26.38	224		28.96	244		31.55	264		34.13	284		36.72
205		26.50	225		29.09	245		31.68	265		34.26	285		36.85
206		26.63	226		29.22	246		31.81	266		34.39	286		36.98
207		26.76	227		29.35	247		31.94	267		34.52	287		37.11
208		26.89	228		29.48	248		32.06	268		34.65	288		37.24
209		27.02	229		29.61	249		32.19	269		34.78	289		37.37
210		27.15	**230**		29.74	**250**		32.32	**270**		34.91	**290**		37.49
211		27.28	231		29.87	251		32.45	271		35.04	291		37.62
212		27.41	232		30.00	252		32.58	272		35.17	292		37.75
213		27.54	233		30.13	253		32.71	273		35.30	293		37.88
214		27.67	234		30.25	254		32.84	274		35.43	294		38.01
215		27.80	235		30.38	255		32.97	275		35.56	295		38.14
216		27.93	236		30.51	256		33.10	276		35.68	296		38.27
217		28.06	237		30.64	257		33.23	277		35.81	297		38.40
218		28.19	238		30.77	258		33.36	278		35.94	298		38.53
219		28.32	239		30.90	259		33.49	279		36.07	299		38.66

BULLION ROOMS, ASSAY OFFICE, CHEMICAL LABORATORY AND ORE FLOORS,

524 Sacramento Street, San Francisco.

TABLE OF THE VALUE OF
SILVER

Per OUNCE TROY, AT DIFFERENT DEGREES OF FINENESS

–BY–

THOMAS PRICE.

FINE	DOLLARS CENTS	FINE	DOLLARS CENTS	FINE	DOLLARS CENTS	FINE	DOLLARS CENTS	FINE	DOLLARS CENTS
300	38.79	**320**	41.37	**340**	43.96	**360**	46.55	**380**	49.13
301	38.92	321	41.50	341	44.09	361	46.67	381	49.26
302	39.05	322	41.63	342	44.22	362	46.80	382	49.39
303	39.18	323	41.76	343	44.35	363	46.93	383	49.52
304	39.30	324	41.89	344	44.48	364	47.06	384	49.65
305	39.43	325	42.02	345	44.61	365	47.19	385	49.78
306	39.56	326	42.15	346	44.74	366	47.32	386	49.91
307	39.69	327	42.28	347	44.86	367	47.45	387	50.04
308	39.82	328	42.41	348	44.99	368	47.58	388	50.17
309	39.95	329	42.54	349	45.12	369	47.71	389	50.29
310	40.08	**330**	42.67	**350**	45.25	**370**	47.84	**390**	50.42
311	40.21	331	42.80	351	45.38	371	47.97	391	50.55
312	40.34	332	42.93	352	45.51	372	48.10	392	50.68
313	40.47	333	43.05	353	45.64	373	48.23	393	50.81
314	40.60	334	43.18	354	45.77	374	48.36	394	50.94
315	40.73	335	43.31	355	45.90	375	48.48	395	51.07
316	40.86	336	43.44	356	46.03	376	48.61	396	51.20
317	40.98	337	43.57	357	46.16	377	48.74	397	51.33
318	41.11	338	43.70	358	46.29	378	48.87	398	51.46
319	41.24	339	43.83	359	46.42	379	49.00	399	51.59

BULLION ROOMS, ASSAY OFFICE, CHEMICAL LABORATORY AND ORE FLOORS,

524 Sacramento Street, San Francisco.

TABLE OF THE VALUE OF
SILVER

Per OUNCE TROY, AT DIFFERENT DEGREES OF FINENESS

–BY–

THOMAS PRICE.

FINE	DOLLARS	CENTS	FINE	DOLLARS	CENTS	FINE	DOLLARS	CENTS	FINE	DOLLARS	CENTS	FINE	DOLLARS	CENTS
400		51.72	420		54.30	440		56.89	460		59.47	480		62.06
401		51.85	421		54.43	441		57.02	461		59.60	481		62.19
402		51.98	422		54.56	442		57.15	462		59.73	482		62.32
403		52.11	423		54.69	443		57.28	463		59.86	483		62.45
404		52.23	424		54.82	444		57.41	464		59.99	484		62.58
405		52.36	425		54.95	445		57.54	465		60.12	485		62.71
406		52.49	426		55.08	446		57.66	466		60.25	486		62.84
407		52.62	427		55.21	447		57.79	467		60.38	487		62.97
408		52.75	428		55.34	448		57.92	468		60.51	488		63.09
409		52.88	429		55.47	449		58.05	469		60.64	489		63.22
410		53.01	430		55.60	450		58.18	470		60.77	490		63.35
411		53.14	431		55.73	451		58.31	471		60.90	491		63.48
412		53.27	432		55.85	452		58.44	472		61.03	492		63.61
413		53.40	433		55.98	453		58.57	473		61.15	493		63.74
414		53.53	434		56.11	454		58.70	474		61.28	494		63.87
415		53.66	435		56.24	455		58.83	475		61.41	495		64.00
416		53.79	436		56.37	456		58.96	476		61.54	496		64.13
417		53.92	437		56.50	457		59.09	477		61.67	497		64.26
418		54.04	438		56.63	458		59.22	478		61.80	498		64.39
419		54.17	439		56.76	459		59.35	479		61.93	499		64.52

BULLION ROOMS, ASSAY OFFICE, CHEMICAL LABORATORY AND ORE FLOORS,

524 Sacramento Street, San Francisco.

TABLE OF THE VALUE OF
SILVER
Per OUNCE TROY, AT DIFFERENT DEGREES OF FINENESS

-BY-

THOMAS PRICE.

FINE.	DOLLARS.	CENTS.	FINE.	DOLLARS.	CENTS.	FINE.	DOLLARS.	CENTS.	FINE.	DOLLARS.	CENTS.	FINE.	DOLLARS.	CENTS.
500		64.65	520		67.23	540		69.82	560		72.40	580		74.99
501		64.78	521		67.36	541		69.95	561		72.53	581		75.12
502		64.91	522		67.49	542		70.08	562		72.66	582		75.25
503		65.03	523		67.62	543		70.21	563		72.79	583		75.38
504		65.16	524		67.75	544		70.34	564		72.92	584		75.51
505		65.29	525		67.88	545		70.46	565		73.05	585		75.64
506		65.42	526		68.01	546		70.59	566		73.18	586		75.77
507		65.55	527		68.14	547		70.72	567		73.31	587		75.89
508		65.68	528		68.27	548		70.85	568		73.44	588		76.02
509		65.81	529		68.40	549		70.98	569		73.56	589		76.15
510		65.94	530		68.53	550		71.11	570		73.69	590		76.28
511		66.07	531		68.65	551		71.24	571		73.82	591		76.41
512		66.20	532		68.78	552		71.37	572		73.95	592		76.54
513		66.33	533		68.91	553		71.50	573		74.08	593		76.67
514		66.46	534		69.04	554		71.63	574		74.21	594		76.80
515		66.59	535		69.17	555		71.76	575		74.34	595		76.93
516		66.72	536		69.30	556		71.89	576		74.47	596		77.06
517		66.84	537		69.43	557		72.02	577		74.60	597		77.19
518		66.97	538		69.56	558		72.15	578		74.73	598		77.32
519		67.10	539		69.69	559		72.27	579		74.86	599		77.45

BULLION ROOMS, ASSAY OFFICE, CHEMICAL LABORATORY AND ORE FLOORS,

524 Sacramento Street, San Francisco.

TABLE OF THE VALUE OF
SILVER
Per OUNCE TROY, AT DIFFERENT DEGREES OF FINENESS
—BY—
THOMAS PRICE.

FINE	DOLLARS	CENTS	FINE	DOLLARS	CENTS	FINE	DOLLARS	CENTS	FINE	DOLLARS	CENTS	FINE	DOLLARS	CENTS
600		77.58	**620**		80.16	**640**		82.75	**660**		85.33	**680**		87.92
601		77.71	621		80.29	641		82.88	661		85.46	681		88.05
602		77.83	622		80.42	642		83.01	662		85.59	682		88.18
603		77.96	623		80.55	643		83.14	663		85.72	683		88.31
604		78.09	624		80.68	644		83.26	664		85.85	684		88.44
605		78.22	625		80.81	645		83.39	665		85.98	685		88.57
606		78.35	626		80.94	646		83.52	666		86.11	686		88.69
607		78.48	627		81.07	647		83.65	667		86.24	687		88.82
608		78.61	628		81.20	648		83.78	668		86.37	688		88.95
609		78.74	629		81.33	649		83.91	669		86.50	689		89.08
610		78.87	**630**		81.45	**650**		84.04	**670**		86.63	**690**		89.21
611		79.00	631		81.58	651		84.17	671		86.76	691		89.34
612		79.13	632		81.71	652		84.30	672		86.88	692		89.47
613		79.26	633		81.84	653		84.43	673		87.01	693		89.60
614		79.39	634		81.97	654		84.56	674		87.14	694		89.73
615		79.52	635		82.10	655		84.69	675		87.27	695		89.86
616		79.64	636		82.23	656		84.82	676		87.40	696		89.99
617		79.77	637		82.36	657		84.95	677		87.53	697		90.12
618		79.90	638		82.49	658		85.07	678		87.66	698		90.25
619		80.03	639		82.62	659		85.20	679		87.79	699		90.38

BULLION ROOMS, ASSAY OFFICE, CHEMICAL LABORATORY AND ORE FLOORS,
524 Sacramento Street, San Francisco.

TABLE OF THE VALUE OF

SILVER

Per OUNCE TROY, AT DIFFERENT DEGREES OF FINENESS

–BY–

THOMAS PRICE.

FINE.	DOLLARS	CENTS.	FINE.	DOLLARS	CENTS.	FINE.	DOLLARS	CENTS.	FINE.	DOLLARS	CENTS.	FINE.	DOLLARS	CENTS.
700		90.51	720		93.09	740		95.68	760		98.26	780	1	00.85
701		90.63	721		93.22	741		95.81	761		98.39	781	1	00.98
702		90.76	722		93.35	742		95.94	762		98.52	782	1	01.11
703		90.89	723		93.48	743		96.06	763		98.65	783	1	01.24
704		91.02	724		93.61	744		96.19	764		98.78	784	1	01.37
705		91.15	725		93.74	745		96.32	765		98.91	785	1	01.49
706		91.28	726		93.87	746		96.45	766		99.04	786	1	01.62
707		91.41	727		94.00	747		96.58	767		99.17	787	1	01.75
708		91.54	728		94.13	748		96.71	768		99.30	788	1	01.88
709		91.67	729		94.25	749		96.84	769		99.43	789	1	02.01
710		91.80	730		94.38	750		96.97	770		99.56	790	1	02.14
711		91.93	731		94.51	751		97.10	771		99.68	791	1	02.27
712		92.06	732		94.64	752		97.23	772		99.81	792	1	02.40
713		92.19	733		94.77	753		97.36	773		99.94	793	1	02.53
714		92.32	734		94.90	754		97.49	774	1	00.07	794	1	02.66
715		92.44	735		95.03	755		97.62	775	1	00.20	795	1	02.79
716		92.57	736		95.16	756		97.75	776	1	00.33	796	1	02.92
717		92.70	737		95.29	757		97.87	777	1	00.46	797	1	03.05
718		92.83	738		95.42	758		98.00	778	1	00.59	798	1	03.18
719		92.96	739		95.55	759		98.13	779	1	00.72	799	1	03.31

BULLION ROOMS, ASSAY OFFICE, CHEMICAL LABORATORY AND ORE FLOORS,

524 Sacramento Street, San Francisco.

TABLE OF THE VALUE OF

SILVER

Per OUNCE TROY, AT DIFFERENT DEGREES OF FINENESS

–BY–

THOMAS PRICE.

FINE	DOLLARS	CENTS	FINE	DOLLARS	CENTS	FINE	DOLLARS	CENTS	FINE	DOLLARS	CENTS	FINE	DOLLARS	CENTS
800	1	03.43	820	1	06.02	840	1	08.61	860	1	11.19	880	1	13.78
801	1	03.56	821	1	06.15	841	1	08.74	861	1	11.32	881	1	13.91
802	1	03.69	822	1	06.28	842	1	08.86	862	1	11.45	882	1	14.04
803	1	03.82	823	1	06.41	843	1	08.99	863	1	11.58	883	1	14.17
804	1	03.95	824	1	06.54	844	1	09.12	864	1	11.71	884	1	14.29
805	1	04.08	825	1	06.67	845	1	09.25	865	1	11.84	885	1	14.42
806	1	04.21	826	1	06.80	846	1	09.38	866	1	11.97	886	1	14.55
807	1	04.34	827	1	06.93	847	1	09.51	867	1	12.10	887	1	14.68
808	1	04.47	828	1	07.05	848	1	09.64	868	1	12.23	888	1	14.81
809	1	04.60	829	1	07.18	849	1	09.77	869	1	12.36	889	1	14.94
810	1	04.73	830	1	07.31	850	1	09.90	870	1	12.48	890	1	15.07
811	1	04.86	831	1	07.44	851	1	10.03	871	1	12.61	891	1	15.20
812	1	04.99	832	1	07.57	852	1	10.16	872	1	12.74	892	1	15.33
813	1	05.12	833	1	07.70	853	1	10.29	873	1	12.87	893	1	15.46
814	1	05.24	834	1	07.83	854	1	10.42	874	1	13.00	894	1	15.59
815	1	05.37	835	1	07.96	855	1	10.55	875	1	13.13	895	1	15.72
816	1	05.50	836	1	08.09	856	1	10.67	876	1	13.26	896	1	15.85
817	1	05.63	837	1	08.22	857	1	10.80	877	1	13.39	897	1	15.98
818	1	05.76	838	1	08.35	858	1	10.93	878	1	13.52	898	1	16.11
819	1	05.89	839	1	08.48	859	1	11.06	879	1	13.65	899	1	16.23

BULLION ROOMS, ASSAY OFFICE, CHEMICAL LABORATORY AND ORE FLOORS,

524 Sacramento Street, San Francisco.

TABLE OF THE VALUE OF
SILVER
Per OUNCE TROY, AT DIFFERENT DEGREES OF FINENESS
-BY-
THOMAS PRICE.

FINE	DOLLARS	CENTS	FINE	DOLLARS	CENTS	FINE	DOLLARS	CENTS	FINE	DOLLARS	CENTS	FINE	DOLLARS	CENTS
900	1	16.36	**920**	1	18.95	**940**	1	21.54	**960**	1	24.12	**980**	1	26.71
901	1	16.49	921	1	19.08	941	1	21.66	961	1	24.25	981	1	26.84
902	1	16.62	922	1	19.21	942	1	21.79	962	1	24.38	982	1	26.97
903	1	16.75	923	1	19.34	943	1	21.92	963	1	24.51	983	1	27.09
904	1	16.88	924	1	19.47	944	1	22.05	964	1	24.64	984	1	27.22
905	1	17.01	925	1	19.60	945	1	22.18	965	1	24.77	985	1	27.35
906	1	17.14	926	1	19.73	946	1	22.31	966	1	24.90	986	1	27.48
907	1	17.27	927	1	19.85	947	1	22.44	967	1	25.03	987	1	27.61
908	1	17.40	928	1	19.98	948	1	22.57	968	1	25.16	988	1	27.74
909	1	17.53	929	1	20.11	949	1	22.70	969	1	25.28	989	1	27.87
910	1	17.66	**930**	1	20.24	**950**	1	22.83	**970**	1	25.41	**990**	1	28.00
911	1	17.79	931	1	20.37	951	1	22.96	971	1	25.54	991	1	28.13
912	1	17.92	932	1	20.50	952	1	23.09	972	1	25.67	992	1	28.26
913	1	18.04	933	1	20.63	953	1	23.22	973	1	25.80	993	1	28.39
914	1	18.17	934	1	20.76	954	1	23.35	974	1	25.93	994	1	28.52
915	1	18.30	935	1	20.89	955	1	23.47	975	1	26.06	995	1	28.65
916	1	18.43	936	1	21.02	956	1	23.60	976	1	26.19	996	1	28.78
917	1	18.56	937	1	21.15	957	1	23.73	977	1	26.32	997	1	28.91
918	1	18.69	938	1	21.28	958	1	23.86	978	1	26.45	998	1	29.03
919	1	18.82	939	1	21.41	959	1	23.99	979	1	26.58	999	1	29.16

1000	1	29.29

BULLION ROOMS, ASSAY OFFICE, CHEMICAL LABORATORY AND ORE FLOORS,
524 Sacramento Street, San Francisco.

RULE FOR PURCHASE AND SALE OF UNPARTED GOLD BARS, AND EXPLANATIONS.

It has been explained in the Preface to the Gold Table, that every $\frac{1}{1000}$ (.001) of gold is worth $0.02.06718, etc., and hence the par rate at which unparted gold bars may at any time sell, simply means the price, and is only the value of so many thousandths of gold. For instance, if 850 is par, the price is $17.57 per oz., because this is the value of eight hundred and fifty thousandths.

The par rate then being simply the price, and the price changing from day to day, with the supply and demand, a necessity arose for some simple rule by which all degrees of fineness could be reduced to any par rate which might, for the moment, prevail. The rule adopted is, for all degrees of fineness above the par rate, deduct from the value of the bar to be sold the $\frac{1}{100}$ of 1 per cent. for every $\frac{1}{1000}$ (.001) difference in fineness, and for all degrees of fineness below the par rate, add to the value of the bar to be sold the $\frac{1}{100}$ of 1 per cent. for every .001 of difference in fineness. The theory of the rule is, that while the fineness and value of the gold alone are stamped upon the bar, yet the silver contained constitutes the profit of the purchaser or covers the cost of converting the gold into coin. For instance, in a bar which is 850 fine in gold, it is assumed that there are 140 of silver, and hence if 850 is par, and the bar to be sold is 890 fine, it must be discounted 4-10 of 1 per cent., because it is deficient .040 in silver as compared with the par rate. But if the bar to be sold should be 810 fine, a premium of 4-10 of 1 per cent. must be allowed, because it contains .040 of silver more than the par rate requires.

The rule, however, is by no means exact, but has remained in force from an early day, when the discrepancies resulting from it were not regarded as of any importance. For instance, in the example above the 4-10 of 1 per cent. on a bar of 100 ozs. 890 fine would be $7.36, whereas on 810 fine it would be $6.69—representing, in both cases, .040 of silver, which are only worth $5.17—being in one case an excess of 42 per cent. and the other of 29 per cent.—a difference of 13 per cent. Indeed, the percentage on the gold is equal to the silver in the single instance where the gold is 625 fine. For all degrees above that fineness the percentage of gold amounts to more than the value of the difference in silver, and to less for all degrees below 625. Under this rule, therefore, the purchaser always gains when he buys bars above the par rate, and loses when he buys those below it.

In New York the rule is different, and approximates nearer the actual differences. There 900 fine is assumed as a fixed basis (instead of having a variable par rate as with us) and the fluctuations in the price of bullion are adjusted by a discount or premium. The theory of this rule is, that 8-10 of the difference between this standard and the fineness of the bullion to be sold represent the actual difference in silver (75-100 would be still nearer.)

EXAMPLE:—Gold bar 100 ozs., 840 fine, worth.............................$1,736 43
(Sold @ $\frac{3}{4}$ premium on 900.)
900—840=60×8=48.0+37$\frac{1}{2}$($\frac{3}{4}$)=85$\frac{1}{2}$c. premium per $100........... 14 84
_____
$1,751 27

85$\frac{1}{2}$c. per $100 equals .085$\frac{1}{4}$ in fineness, plus 840 ÷ 925$\frac{1}{4}$ par, as per San Francisco rule.

GOLD COINS.

COUNTRY.	DENOMINATIONS.	WEIGHT.	FINENESS.	VALUE.	VALUE AFTER DEDUCTION.
		Oz. Dec.	Thousandths.		
Australia	Pound of 1852	0.281	916.5	$5.32.37	$5.29.71
Australia	Sovereign of 1855–60	0.256.5	916	4.85.58	4.83.16
Austria	Ducat	0.112	986	2.28.28	2.27.04
Austria	Souverain	0.363	900	6.75.35	6.71.98
Austria	New Union Crown (assumed)	0.357	900	6.64.19	6.60.87
Belgium	Twenty-five francs	0.254	899	4.72.03	4.69.67
Bolivia	Doubloon	0.867	870	15.59.25	15.51.46
Brazil	20 milreis	0.575	917.5	10.90.57	10.85.12
Central America	Two escudos	0.209	853.5	3.68.75	3.66.91
Central America	Four reals	0.027	875	0.48.8	0.48.6
Chile	Old Doubloon	0.867	870	15.59.26	15.51.47
Chile	Ten Pesos	0.492	900	9.15.35	9.10.78
Denmark	Ten thaler	0.427	895	7.90.01	7.86.06
Ecuador	Four escudos	0.433	844	7.55.46	7.51.69
England	Pound or Sovereign, new	0.256.7	916.5	4.86.34	4.83.91
England	Pound or Sovereign, average	0.256.2	916	4.84.91	4.82.50
France	Twenty francs, new	0.207.5	899.5	3.85.83	3.83.91
France	Twenty francs, average	0.207	899	3.84.69	3.82.77
Germany, North	Ten thaler	0.427	895	7.90.01	7.86.06
Germany, North	Ten thaler, Prussian	0.427	903	7.97.07	7.93.09
Germany, North	Krone (crown)	0.357	900	6.64.20	6.60.88
Germany, South	Ducat	0.112	986	2.28.28	2.27.14
Greece	Twenty drachms	0.185	900	3.44.19	3.42.47
Hindostan	Mohur	0.374	916	7.08.18	7.04.64
Italy	Twenty lire	0.207	898	3.84.26	3.82.34
Japan	Old cobang	0.362	568	4.44.0	4.41.8
Japan	New cobang	0.289	572	3.57.6	3.55.8
Mexico	Doubloon, average	0.867.5	866	15.52.98	15.45.22
Mexico	Doubloon, new	0.867.5	870.5	15.61.05	15.53.25
Naples	Six Ducati, new	0.245	996	5.04.43	5.01.91
Netherland	Ten Guilders	0.215	899	3.99.56	3.97.57
New Granada	Old Doubloon, Bogota	0.868	870	15.61.06	15.53.26
New Granada	Old Doubloon, Popayan	0.867	858	15.37.75	15.30.07
New Granada	Ten pesos, new	0.525	891.5	9.67.51	9.62.68
Peru	Old Doubloon	0.867	868	15.55.67	15.47.90
Peru	Twenty soles	1.035	898	19.21.8	19.12.2
Portugal	Gold Crown	0.308	912	5.80.66	5.77.76
Prussia	New Union Crown (assumed)	0.357	900	6.64.19	6.60.87
Rome	2½ scudi, new	0.140	900	2.60.47	2.59.17
Russia	Five roubles	0.210	916	3.97.64	3.95.66
Spain	100 reals	0.268	896	4.96.39	4.93.91
Spain	80 reals	0.215	869.5	3.86.44	3.84.51
Sweden	Ducat	0.111	975	2.23.72	2.22.61
Tunis	25 piastres	0.161	900	2.99.54	2.98.05
Turkey	100 piastres	0.231	915	4.36.93	4.34.75
Tuscany	Sequin	0.112	999	2.31.29	2.30.14

SILVER COINS.

COUNTRY.	DENOMINATIONS.	WEIGHT.	FINENESS.	VALUE.
		Oz. Dec.	Thousandths.	
Austria	Old rix dollar	0.902	833	$1.02.27
Austria	Old scudo	0.836	902	1.02.64
Austria	Florin before 1858	0.451	833	51.14
Austria	New florin	0.397	900	48.63
Austria	New Union dollar	0.596	900	73.01
Austria	Maria Theresa dollar, 1780	0.895	838	1.02.12
Belgium	Five francs	0.803	897	98.04
Bolivia	New dollar	0.643	903.5	79.07
Bolivia	Half dollar	0.432	667	39.22
Brazil	Double milreis	0.820	918.5	1.02.53
Canada	20 cents	0.150	925	18.87
Central America	Dollar	0.866	850	1.00.19
Chile	Old dollar	0.864	908	1.06.79
Chile	New dollar	0.801	900.5	98.17
Denmark	Two rigsdaler	0.927	877	1.10.65
England	Shilling, new	0.182.5	924.5	22.96
England	Shilling, average	0.178	925	22.41
France	Five franc, average	0.800	900	98.00
Germany, North	Thaler, before 1857	0.712	750	72.67
Germany, North	New thaler	0.595	900	72.89
Germany, South	Florin, before 1857	0.340	900	41.65
Germany, South	New florin (assumed)	0.340	900	41.65
Greece	Five drachms	0.719	900	88.08
Hindostan	Rupee	0.374	916	46.62
Japan	Itzebu	0.279	991	37.63
Japan	New Itzebu	0.279	890	33.80
Mexico	Dollar, new	0.867.5	903	1.06.62
Mexico	Dollar, average	0.866	901	1.06.20
Naples	Scudo	0.844	830	95.34
Netherlands	2½ guild	0.804	944	1.03.31
Norway	Specie daler	0.927	877	1.10.65
New Granada	Dollar of 1857	0.803	896	97.92
Peru	Old dollar	0.866	901	1.06.20
Peru	Dollar of 1858	0.766	909	94.77
Peru	Half-dollar, 1835-38	0.433	650	38.31
Prussia	Thaler, before 1857	0.712	750	72.68
Prussia	New thaler	0.595	900	72.89
Rome	Scudo	0.864	900	1.05.84
Russia	Rouble	0.667	875	79.44
Sardinia	Five lire	0.800	900	98.00
Spain	New pistareen	0.166	899	20.31
Sweden	Rix dollar	1.092	750	1.11.48
Switzerland	Two francs	0.323	899	39.52
Tunis	Five piastres	0.511	898.5	62.49
Turkey	Twenty piastres	0.770	830	86.98
Tuscany	Florin	0.220	925	27.60

EXPLANATION OF ENGLISH RETURNS.

The English standard for gold is 22 carats of 4 grains each, equal to 88 carat grains. Their pure is 24 carats, or 96 grains.

24 carats is equal to 1000.   decimal fineness.

1   "   "   .041.6666   "   "

1 carat grain equal .010.4166   "   "

$\frac{1}{8}$ "   "   "   .001.302083 "   "

Hence, 24 carats or 96 grains pure corresponding with 1000, to express their standard in decimal fineness, multiply 88 by 1000, and divide by 96, equals 916$\frac{2}{3}$, our fineness.

The English standard for silver is 222 dwts.   Their pure is 240 dwts.

12 ozs. or 240 dwts. is equal to 1000.   decimal fineness.

1   "   20   "   "   .083.3333   "   "

1 dwt. is equal to   .004.1666   "   "

Hence, to convert their standard into decimal fineness, multiply 222 by 1000 and divide by 240, equals .925.

In all English assayers' reports of metals they state the *betterness* or *worseness*, as compared with their standard, and to ascertain the standard weight of metal reported, the gross weight, in case of gold multiplied by the sum after adding to 88 for *better* or deducting from 88 for *worse*, as the case may be, and dividing this product by 88, the result is standard ounces, thus:

560.21 ozs. gross weight gold, reported W.6$\frac{1}{4}$=520.42 standard ounces.

(6$\frac{1}{4}$ from 88 equal 81$\frac{3}{4}$×560.21÷88=520.42.)

The equivalent of 6$\frac{1}{4}$ worse than English standard expressed in decimal fineness is determined, viz: 81$\frac{3}{4}$×1000÷96=851$\frac{1}{2}$. If the report had been 6$\frac{1}{4}$ better, then add to 88=94$\frac{1}{4}$×560.21÷88=599.99 standard ounces. Expressed in decimal fineness, 6$\frac{1}{4}$ better= 94$\frac{1}{4}$×1000÷96=981.7.

Or thus, for silver:

450.12 ozs. gross, silver, reported W. 15 dwts.=419.706 standard ounces, 222—15=207. ×450.12÷222=419.706. Expressed in decimal fineness, 207.×1000÷240=862$\frac{1}{2}$. A shorter rule for silver equivalent is to multiply better or worse by 4 1-6, and add or deduct the product from .925, English standard, as the case may be. In the above case 15×4 1-6=62.5 from 925= 862..5

CHARGES FOR REFINING.

In England the charges are met by the refiner becoming the purchaser of the metals to be refined and the seller of the metals so refined. The difference in price and deductions includes the cost of working and the refiner's profit.

For refining unparted gold bars, the London refiners deduct 20 dwts. of the silver contained per lb. Troy of the metal, which, at 5s. 6d. per oz., (the price allowed for silver over 20 dwts.) amounts to 11.1c. per oz. gross. They usually allow, however, 1d. premium on the gold, which practically reduces the charge to 9¼c. per oz. gross. But it has been found, by experience, that when the charges for assaying, melting and other incidentals customary are added, the aggregate of the charges will average nearly 12c. per oz. gross.

For refining Silver Doré, the deduction is 5 grs. of fine gold for every lb. Troy of the metal, equal to 1.79c. per oz. gross. They then allow 84s. 7¼d. for the remaining fine gold =$20.58.62 per oz. instead of $20.67.18., the true value—and with the difference in the price of the gold, together with the charges for assaying, melting and other extra expenses customary, it is found that the charges are practically 2.03c. per oz. gross. But in cases where the proportion of gold contained is greater than 5 grs. per lb. of the total metal, the charges increase in the same proportion, because of the reduced price allowed for the fine gold, and when the gold contained is equal to 1000 grs. per lb. of the metal, corresponding with .173 in decimal fineness, the charge is 2½d. or 5c. per oz.

In cases where the gold contained is not equal to 5 grs. per lb. gross, then a deduction is made on the price of the silver, which varies with the proportion of gold or base metal present. This difference is from ½ to 1½d. per standard oz.

The Silver table following will illustrate the application of the above, which is based only on the charge of 5 grs. per lb. and assay charge, yet in continuing the calculation to silver containing .170 parts gold, it will be seen that the charge in this case equals 3.36c. per oz.

The calculations of the following Tables are based upon the *intrinsic* value of the English pound, viz: $4.86.65, and of the penny 2.02.77c. (the pound contains legally 113.001 grains of fine gold, which, at 4.3066c., the value of one grain, equals $4.86.65.)

TABLE

Showing Percentage of Discount or Premium on the value of 100 ozs. Gold Bullion at different degrees of Fineness, based on the out turn in London at the following rates per standard oz. for the Gold, and the Silver contained at 5s. 6d. per fine oz., deducting as the *approximated* average charge for refining 12c. per oz. gross, in lieu of the 20 dwts. silver per lb. Troy deducted and the charge for assay.

FINENESS. Gold.	Premium or Discount per $100 on out turn at 77s. 10d.	Premium per $100 on out turn at 77s. 10¼d.	Premium per $100 on out turn at 77s. 11d.	Premium per $100 on out turn at 77s. 11¼d.	Premium per $100 on out turn at 78s.
900	Dis. .05.3	Par.	.05.3	.10.5	.16.
895	" .01.7	.03.6	.08.9	.14.3	.19.6
890	Prem. .02.	.07.3	.12.7	.18.	.23.3
885	.04.6	.10.9	.16.3	.21.6	.27.
880	.09.4	.14.7	.20.1	.25.4	.30.7
875	.13.2	.18.5	.23.8	.29.1	.34.5
870	.17.	.22.3	.27.7	.33.	.38.3
865	.20.	.26.3	.31.7	.37.	.42.3
860	.24.8	.30.2	.35.4	.40.8	.46.2
855	.28.7	.34.	.39.3	.44.7	.49.8
850	.32.7	.38.	.43.4	.48.7	.54.
845	.36.7	.42.1	.47.4	.52.7	.58.1
840	.41.	.46.3	.51.6	.56.9	.62.3
835	.45.2	.51.	.55.7	.61.1	.66.4
830	.49.3	.54.6	.59.9	.65.3	.70.6
825	.53.5	.58.9	.64.6	.69.6	.74.9
820	.57.8	.63.2	.68.5	.73.8	.79.1
815	.62.2	.67.6	.72.9	.78.3	.83.6
810	.66.7	.72.	.77.3	.82.7	.88.
805	.71.2	.76.4	.81.7	.87.1	.92.4
800	.75.5	.80.9	.86.2	.91.5	.96.8
795	.80.2	.85.4	.90.9	.96.2	1.01.5

EXAMPLE.—Gold Bar of 100 ozs. at 900 fine, value of gold.......................$1,860.46

90 ozs. fine gold = 98.181 English standard at 77s 10d. ($18.93.87)..$1,859.42

9 ozs. fine silver, at 5s. 6d. ($1.33.82)........................ 12.04

$1,871.46

Deduct 12c. per oz. (in lieu of 20 dwts. and assay charge) for refining 12.00

Net out turn.. $1,859.46

Loss... $1.00

$1 00 ÷ 1860.46 = ₅₃/₁₀₀₀π per cent. premium as above.

TABLE

Giving Percentage of Discount or Premium on Total Value of 1000 ozs. Silver Bullion at different Degrees of Fineness, based on the out turn in London at the following rates per standard oz. for the Silver, and for the Gold contained 84s. 7¼d.=$20.58.62 per fine oz., estimating the charges at 5 grs. Gold deducted per lb. Troy of metal, and assaying at 5s.

FINENESS		Prem. or Discount on out turn at 60½ d. per st'd oz. ($1.22.67.)	Prem. or Discount on out turn at 60¾d. per st'd oz ($1.23.18.)	Prem. on out turn at 61d. per standard ounce. ($1.23.60.)	Prem. on out turn at 61¼d. per standard ounce. ($1.24.19.)	Prem. on out turn at 61½d. per standard ounce ($1.24.69.)	Prem. on out turn at 61¾d. per standard ounce. ($1.25.20.)	Prem. on out turn at 62d. per standard ounce. ($1.25.71.)	Cost per oz for Ref'ng. (See explanations.)
Silver.	Gold.								
980	10	.00.86c.	01.22c.	.01.59c.	.01.95c.	.02.31c.	.02.67c.	.03.04c.	1.99c.
970	20	.68	1.01	1.33	1.64	1.95	2.28	2.60	2.08
960	30	.55	.83	1.12	1.40	1.68	1.96	2.25	2.16
950	40	.44	.69	.95	1.20	1.45	1.70	1.96	2.25
940	50	.35	.58	.81	1.04	1.26	1.49	1.72	2.34
930	60	.27	.48	.69	.90	1.10	1.31	1.52	2.42
920	70	.21	.40	.59	.78	.97	1.16	1.36	2.51
910	80	.15	.33	.51	.68	.85	1.03	1.21	2.59
900	90	.10	.27	.43	.59	.75	.92	1.08	2.68
890	100	.06	.21	.36	51	.66	.82	.97	2.76
880	110	.02	.16	.31	.45	.58	.73	.87	2.85
870	120	dist. .01	.12	.25	.38	.51	.65	.78	2.94
860	130	.04	.08	.21	.33	.45	.57	.70	3.02
850	140	.07	.05	.16	.28	.39	.51	.63	3.11
840	150	.10	.01	.13	.23	.34	.45	.56	3.19
830	160	.12	dis't. 01	.09	.19	.29	.40	.50	3.28
820	170	.14	.04	.06	.15	.25	.35	.45	3.36

EXAMPLE.—Silver Bar 1000 ozs. gross, 980 fine silver, allowing 10 parts for base metal, 10 fine gold. Total value................................... $1,473 79

1000 ozs. 980 fine=980 ozs. fine silver ÷ 925=1059.45 ozs. English stan'd silver at 60½d. ($1.22.67.)..$1,299 63

10 ozs. fine gold, less 5 grains per lb. gross=83⅓ lb.=416⅔ grains÷480= .868 ozs., which from 10 ozs.=9.132 ozs. at 84s. 7¼d. ($20.58.62).... 187 99

$1,487 62

Less one assay... 1 22

Net out turn... $1,486 40

Profit............................... $12 61

12.61÷1473.79=$\frac{86}{100}$ per cent. premium as above.

Tables Prepared by Thomas Price, 524 Sacramento Street, S. F.

Decimal Fineness Expressed in Equivalents of Betterness or Worseness than English Standard for GOLD.

Fineness in 1000ths.	Equivalent in English reports better or worse than standard .88 grns	Fineness in 1000ths.	Equivalent in English reports better or worse than standard .88 grns	Fineness in 1000ths.	Equivalent in English reports better or worse than standard .88 grns	Fineness in 1000ths.	Equivalent in English reports better or worse than standard .88 grns	Fineness in 1000ths.	Equivalent in English reports better or worse than standard .88 grns	Fineness in 1000ths.	Equivalent in English reports better or worse than standard .88 grns
1000	B.8.000	977	B.5.792	954	B.3.584	931	B.1.376	908	W. .832	885	W.3.040	862	W.5.248
999	7.904	976	5.696	953	3.488	930	1.280	907	.928	884	3.136	861	5.344
998	7.808	975	5.600	952	3.392	929	1.184	906	1.024	883	3.232	860	5.440
997	7.712	974	5.504	951	3.296	928	1.088	905	1.120	882	3.328	859	5.536
996	7.616	973	5.408	950	3.200	927	0.992	904	1.216	881	3.424	858	5.632
995	7.520	972	5.312	949	3.104	926	0.896	903	1.312	880	3.520	857	5.728
994	7.424	971	5.216	948	3.008	925	0.800	902	1.408	879	3.616	856	5.824
993	7.328	970	5.120	947	2.912	924	0.704	901	1.504	878	3.712	855	5.920
992	7.232	969	5.024	946	2.816	923	0.608	900	1.600	877	3.808	854	6.016
991	7.136	968	4.928	945	2.720	922	0.512	899	1.696	876	3.904	853	6.112
990	7.040	967	4.832	944	2.624	921	0.416	898	1.792	875	4.000	852	6.208
989	6.944	966	4.736	943	2.528	920	0.320	897	1.888	874	4.096	851	6.304
988	6.848	965	4.640	942	2.432	919	0.224	896	1.984	873	4.192	850	6.400
987	6.752	964	4.544	941	2.336	918	0.128	895	2.080	872	4.288	849	6.496
986	6.656	963	4.448	940	2.240	917	0.032	894	2.176	871	4.384	848	6.592
							916 Stan'd						
985	6.560	962	4.352	939	2.144	916	W..064	893	2.272	870	4.480	847	6.688
984	6.464	961	4.256	938	2.048	915	.160	892	2.368	869	4.576	846	6.784
983	6.368	960	4.160	937	1.952	914	.256	891	2.464	868	4.672	845	6.880
982	6.272	959	4.064	936	1.856	913	.352	890	2.560	867	4.768	844	6.976
981	6.176	958	3.968	935	1.760	912	.448	889	2.656	866	4.864	843	7.072
980	6.080	957	3.872	934	1.664	911	.544	888	2.752	865	4.960	842	7.168
979	5.984	956	3.776	933	1.568	910	.640	887	2.848	864	5.056	841	7.264
978	5.888	955	3.680	932	1.472	909	.736	886	2.944	863	5.152	840	7.360

Decimal Fineness Expressed in Equivalents of Betterness or Worseness than English Standard for SILVER.

Fineness in 1000ths.	Equivalent in English reports better or worse than their stan'd —222 dwts.	Fineness in 1000ths.	Equivalent in English reports better or worse than their stan'd —222 dwts.	Fineness in 1000ths.	Equivalent in English reports better or worse than their stan'd —222 dwts.	Fineness in 1000ths.	Equivalent in English reports better or worse than their stan'd —222 dwts.	Fineness in 1000ths.	Equivalent in English reports better or worse than their stan'd —222 dwts.	Fineness in 1000ths.	Equivalent in English reports better or worse than their stan'd —222 wts.	Fineness in 1000ths.	Equivalent in English reports better or worse than their stan'd —222 dwts.
1000	B.18.	977	B.12.48	954	B. 6.9C	931	B. 1.44	908	Wo.4.08	885	Wo.9.60	862	W.15.12
999	17.76	976	12.24	953	6.72	930	1.20	907	4.32	884	9.84	861	15.36
998	17.52	975	12.00	952	6.48	929	.96	906	4.56	883	10.08	860	15.60
997	17.28	974	11.76	951	6.24	928	.72	905	4.80	882	10.32	859	15.84
996	17.04	973	11.52	950	6.00	927	.48	904	5.04	881	10.56	858	16.08
995	16.80	972	11.28	949	5.76	926	.24	903	5.28	880	10.80	857	16.32
994	16.56	971	11.04	948	5.52	925	Standard.	902	5.52	879	11.04	856	16.56
993	16.32	970	10.80	947	5.28	924	Wo. .24	901	5.76	878	11.28	855	16.80
992	16.08	969	10.56	946	5.04	923	.48	900	6.00	877	11.52	854	17.04
991	15.84	968	10.32	945	4.80	922	.72	899	6.24	876	11.76	853	17.28
990	15.60	967	10.08	944	4.56	921	.96	898	6.48	875	12.00	852	17.52
989	15.36	966	9.84	943	4.32	920	1.20	897	6.72	874	12.24	851	17.76
988	15.12	965	9.60	942	4.08	919	1.44	896	6.96	873	12.48	850	18.00
987	14.88	964	9.36	941	3.84	918	1.68	895	7.20	872	12.72	849	18.24
986	14.64	963	9.12	940	3.60	917	1.92	894	7.44	871	12.96	848	18.48
985	14.40	962	8.88	939	3.36	916	2.16	893	7.68	870	13.20	847	18.72
984	14.16	961	8.64	938	3.12	915	2.40	892	7.92	869	13.44	846	18.96
983	13.92	960	8.40	937	2.88	914	2.64	891	8.16	868	13.68	845	19.20
982	13.68	959	8.16	936	2.64	913	2.88	890	8.40	867	13.92	844	19.44
981	13.44	958	7.92	935	2.40	912	3.12	889	8.64	866	14.16	843	19.68
980	13.20	957	7.68	934	2.16	911	3.36	888	8.88	865	14.40	842	19.92
979	12.96	956	7.44	933	1.92	910	3.60	887	9.12	864	14.64	841	20.16
978	12.72	955	7.20	932	1.68	909	3.84	886	9.36	863	14.88	840	20.40

ENGLISH ASSAY REPORT OF GOLD,

Made in Carats, Carat Grains and Eighths, Better or Worse, expressed in Decimal Fineness—
22 Carats, or 88 Carat Grains, equivalent to 916⅔, English Standard.

Carats.	Carat Grains.	Eighths.	Expressed in 1000ths.	Carats.	Carat Grains.	Eighths.	Expressed in 1000ths.	Carats.	Carat Grains.	Eighths.	Expressed in 1000ths.	Carats.	Carat Grains.	Eighths.	Expressed in 1000ths.	Carats.	Carat Grains.	Eighths.	Expressed in 1000ths.	Carats.	Carat Grains.	Eighths.	Expressed in 1000ths.
		Wo rse.				Wo rse.				Wo rse.				Wo rse.		St and ard			916.7	1	0	0	958.3
4	0	0	750.	3	0	0	791.7	2	0	0	833.3	1	0	0	875.0			Bet ter.		1	0	⅛	959.6
3	3	⅞	751.3	2	3	⅞	793.0	1	3	⅞	834.6	0	3	⅞	876.3	0	0	⅛	918.0	1	0	¼	960.9
3	3	¾	752.6	2	3	¾	794.3	1	3	¾	835.9	0	3	¾	877.6	0	0	¼	919.3	1	0	⅜	962.2
3	3	⅝	753.9	2	3	⅝	795.6	1	3	⅝	837.2	0	3	⅝	878.9	0	0	⅜	920.6	1	0	½	963.5
3	3	½	755.2	2	3	½	796.9	1	3	½	838.5	0	3	½	880.2	0	0	½	921.9	1	0	⅝	964.8
3	3	⅜	756.5	2	3	⅜	798.2	1	3	⅜	839.8	0	3	⅜	881.5	0	0	⅝	923.2	1	0	¾	966.1
3	3	¼	757.8	2	3	¼	799.5	1	3	¼	841.1	0	3	¼	882.8	0	0	¾	924.5	1	0	⅞	967.4
3	3	⅛	759.1	2	3	⅛	800.8	1	3	⅛	842.4	0	3	⅛	884.1	0	0	⅞	925.8	1	1	0	968.8
3	3	0	760.4	2	3	0	802.1	1	3	0	843.8	0	3	0	885.4	0	0	⅞	925.8	1	1	0	968.8
3	2	⅞	761.7	2	2	⅞	803.4	1	2	⅞	845.1	0	2	⅞	886.7	0	1	0	927.1	1	1	⅛	970.1
3	2	¾	763.0	2	2	¾	804.7	1	2	¾	846.4	0	2	¾	888.0	0	1	⅛	928.4	1	1	¼	971.4
3	2	⅝	764.3	2	2	⅝	806.0	1	2	⅝	847.7	0	2	⅝	889.3	0	1	¼	929.7	1	1	⅜	972.7
3	2	½	765.6	2	2	½	807.3	1	2	½	849.0	0	2	½	890.6	0	1	⅜	931.0	1	1	½	974.0
3	2	⅜	766.9	2	2	⅜	808.6	1	2	⅜	850.3	0	2	⅜	891.9	0	1	½	932.3	1	1	⅝	975.3
3	2	¼	768.2	2	2	¼	809.9	1	2	¼	851.6	0	2	¼	893.2	0	1	⅝	933.6	1	1	¾	976.6
3	2	⅛	769.5	2	2	⅛	811.2	1	2	⅛	852.9	0	2	⅛	894.5	0	1	¾	934.9	1	1	⅞	977.9
3	2	0	770.8	2	2	0	812.5	1	2	0	854.2	0	2	0	895.8	0	1	⅞	936.2	1	2	0	979.2
3	1	⅞	772.1	2	1	⅞	813.8	1	1	⅞	855.5	0	1	⅞	897.1	0	2	0	937.5	1	2	⅛	980.5
3	1	¾	773.4	2	1	¾	815.1	1	1	¾	856.8	0	1	¾	898.4	0	2	⅛	938.8	1	2	¼	981.8
3	1	⅝	774.7	2	1	⅝	816.4	1	1	⅝	858.1	0	1	⅝	899.7	0	2	¼	940.1	1	2	⅜	983.1
3	1	½	776.0	2	1	½	817.7	1	1	½	859.4	0	1	½	901.0	0	2	⅜	940.4	1	2	½	984.4
3	1	⅜	777.3	2	1	⅜	819.0	1	1	⅜	860.7	0	1	⅜	902.3	0	2	½	942.7	1	2	⅝	985.7
3	1	¼	778.6	2	1	¼	820.3	1	1	¼	862.0	0	1	¼	903.6	0	2	⅝	944.0	1	2	¾	987.0
3	1	⅛	779.9	2	1	⅛	821.6	1	1	⅛	863.3	0	1	⅛	904.9	0	2	¾	945.3	1	2	⅞	988.3
3	1	0	781.3	2	1	0	822.9	1	1	0	864.6	0	1	0	906.3	0	2	⅞	946.6	1	3	0	989.6
3	0	⅞	782.6	2	0	⅞	824.2	1	0	⅞	865.9	0	0	⅞	907.6	0	3	0	947.9	1	3	⅛	990.9
3	0	¾	783.9	2	0	¾	825.5	1	0	¾	867.2	0	0	¾	908.9	0	3	⅛	949.2	1	3	¼	992.2
3	0	⅝	785.2	2	0	⅝	826.8	1	0	⅝	868.5	0	0	⅝	910.2	0	3	¼	950.5	1	3	⅜	993.5
3	0	½	786.5	2	0	½	828.1	1	0	½	869.8	0	0	½	911.5	0	3	⅜	951.0	1	3	½	994.8
3	0	⅜	787.8	2	0	⅜	829.4	1	0	⅜	871.1	0	0	⅜	912.8	0	3	½	953.1	1	3	⅝	996.1
3	0	¼	789.1	2	0	¼	830.7	1	0	¼	872.4	0	0	¼	914.1	0	3	⅝	954.4	1	3	¾	997.4
3	0	⅛	790.4	2	0	⅛	832.0	1	0	⅛	873.7	0	0	⅛	915.4	0	3	¾	955.7	1	3	⅞	998.7
																0	3	⅞	957.0	2	0	Pu re.	1000.

ENGLISH ASSAY REPORT OF SILVER,

Made in Ozs. and Dwts., Better or Worse, expressed in Decimal Fineness—222 Dwts. per Lb., equivalent to 925, English Standard.

Ounces	Dwts.	Expressed in 1000ths.	Ounces	Dwts.	Expressed in 1000ths.	Ounces	Dwts.	Expressed in 1000ths.	Ounces	Dwts.	Expressed in 1000ths.	Ounces	Dwts.	Expressed in 1000ths.	Ounces	Dwts.	Expressed in 1000ths.	Ounces	Dwts.	Expressed in 1000ths.
Worse.			Worse			Worse.			Worse.			Worse.			Standard.			Better.		
2	10	716.6	2	0	758.3	1	10	800.0	1	0	841.7	0	10	883.3			925.0	0	9½	964.6
2	9½	718.7	1	19½	760.4	1	9½	802.1	0	19½	843.8	0	9½	885.4		Better.		0	10	966.7
2	9	720.8	1	19	762.5	1	9	804.2	0	19	845.8	0	9	887.5	0	0½	927.1	0	10½	968.8
2	8½	722.9	1	18½	764.6	1	8½	806.3	0	18½	847.9	0	8½	889.6	0	1	929.2	0	11	970.8
2	8	725.0	1	18	766.7	1	8	808.3	0	18	850.0	0	8	891.7	0	1½	931.3	0	11½	972.9
2	7½	727.0	1	17½	768.8	1	7½	810.4	0	17½	852.1	0	7½	893.8	0	2	933.3	0	12	975.0
2	7	729.1	1	17	770.8	1	7	812.5	0	17	854.2	0	7	895.8	0	2½	935.4	0	12½	977.1
2	6½	731.2	1	16½	772.9	1	6½	814.6	0	16½	856.3	0	6½	897.9	0	3	937.5	0	13	979.2
2	6	733.3	1	16	775.0	1	6	816.7	0	16	858.3	0	6	900.0	0	3½	939.6	0	13½	981.3
2	5½	735.4	1	15½	777.1	1	5½	818.8	0	15½	860.4	0	5½	902.1	0	4	941.7	0	14	983.3
2	5	737.5	1	15	779.2	1	5	820.8	0	15	862.5	0	5	904.2	0	4½	943.8	0	14½	985.4
2	4½	739.5	1	14½	781.3	1	4½	822.9	0	14½	864.6	0	4½	906.3	0	5	945.8	0	15	987.5
2	4	741.6	1	14	783.3	1	4	825.0	0	14	866.7	0	4	908.3	0	5½	947.9	0	15½	989.6
2	3½	743.7	1	13½	785.4	1	3½	827.1	0	13½	868.8	0	3½	910.4	0	6	950.0	0	16	991.7
2	3	745.8	1	13	787.5	1	3	829.2	0	13	870.8	0	3	912.5	0	6½	952.1	0	16½	993.8
2	2½	747.9	1	12½	789.6	1	2½	831.3	0	12½	872.9	0	2½	914.6	0	7	954.2	0	17	995.8
2	2	750.0	1	12	791.7	1	2	833.3	0	12	875.0	0	2	916.7	0	7½	956.3	0	17½	997.9
2	1½	752.0	1	11½	793.8	1	1½	835.4	0	11½	877.1	0	1½	918.8	0	8	958.3	Pure. 18		1000.0
2	1	754.1	1	11	795.8	1	1	837.5	0	11	879.2	0	1	920.8	0	8½	960.4			
2	0½	756.2	1	10½	797.9	1	0½	839.6	0	10½	881.3	0	0½	922.9	0	9	962.5			

FRENCH WEIGHTS.

Milligramme, 1-1000 of a gramme.........Equal....0.0154 grains.
Centigramme, 1-100 of a gramme.........   "   ....0.1544   "
Décigramme, 1-10 of a gramme............   "   ....1.5440   "
*Gramme (unit of weight)................   "   ..15.44   "
Decagramme, 10 grammes.................   "   ..154.4   "
Hectogramme, 100 grammes..............   "   1544.0   "   { 3.2167 ozs. Troy.
                                                              { 3.5291 ozs. Avd'ps.
Kilogramme 1,000 grammes...............   "   32½ ozs. Troy or 2.2057 lbs. Avd'ps.
Myriagramme, 10,000 grammes...........   "   321⅞ "   "   22.057   "   "

*Some authorities give the gramme as 15.43402344 grains, at which rate the kilogramme would be 32.154 ozs. Troy.

AVOIRDUPOIS CONVERTED INTO METRIC.

1 drm. converted into kilogramme.......................... 0.0017
1 oz.   "   "   " 0.02835
1 lb. (16 ozs.) converted into kilogramme.... .... 0.45359
10 " 4.53593
100 " 45.3593
1000 " 453.59
2000 " 907.19
1 cwt (112 lbs.)....:...................................... 50.80
20 " (2240 lbs.)....................................... 1016.05

METRIC CONVERTED INTO AVOIRDUPOIS.

1 gramme converted into lb. avoirdupois.................... 0.0022046
10 gramme   "   "   " 0.022046
100 grammes...................................... .... 0.220462
1 kilogramme (1000 grammes)........................... 2.20462
10   "   (decagramme)............................. 22.04621
100   "   (Quintal)................................ 220.46213
1000   "   (Millier).................................2204.62125

TROY CONVERTED INTO METRIC.

1 oz. converted into kilogrammes............................ 0.031103
10 "   "   "   " 0.311035
100 "   "   "   " 3.11035
1000 "   "   "   " 31.1035

METRIC CONVERTED INTO TROY OZ.

1 gramme converted into ounces.......................... 0.03215
10 grammes   "   "   " 0.32151
100 grammes   "   "   " 3.21507
1000 grammes—1 kilogramme................................ 32.15073
10 kilogrammes... 321.50727
100   " 3215.0727
1000   " 32150.727

WEIGHTS.

1 gram	=15.432	grains.
	= .03215	oz. Troy
	= .03527	oz. Avoi.
1 grain	= .0648	grms.
1 oz. Troy	=31.103496	grms.
	= 1.0971	oz. Av'r
	= 480	grains.

1 oz. Avoir	= 28.3496	grams.
	= .91145	oz. Troy
	=437.5	grains.

1 cub. in. distilled water (62° F. bar. 30 in.) weighs 252.458 grains.

LINEAR MEASURE.

1 inch =	.0254	metre=2.54 centimetres.
1 metre=39.37079	inches.	
=	1.093569	yards.
=	3.2807	feet.

1 foot =	.30481	metre.
1 yard =	.91443	metre.
1 mile =	1.6093	kilometre.
1 kil'me=	.621383	mile.

MEASURES OF CAPACITY.

1 cu. inch =16.386176 cu. centimetres.
= .004329 U. S. gallon.
= .0036 Imperial "
= .57695 fluid ounce.
1 cu. foot = 7.48 U. S. gallon.
= 6.2321 Imperial gallon.
=28.3177 litre.
1 litre =61.0279 cu. in.
= .03531 cu. ft.
= .264165 U. S. gal.
= .22 Imperial gal.
=35.21 fluid ounces.
1 fluid ounce=28.3962 cu. centimetres =
437.5 grains.
=1.7332 cu. in.

1 U. S. gal'n= 231 cu. in.
= .13368 cu. ft.
= .8331 Imperial gallon.
= 3.7852 litres.
= 8.38888 lbs. Avoir. distilled
water at 39°83 F. Bar. 30 in.

1 Imperial gallon=277.274 cu. in.
= .16046 cu. ft.
= 1.20032 U. S. gal.
= 4.5434 litres.
=10 lb. Avoir. dist. water at
62° F. bar. 30 in.

MEXICAN WEIGHTS.

12 granos	make 1 tomin	Equal..	9.2 grains Troy.	
3 tomines	make 1 adarme	" ..	27.7 grains,	"
2 adarmes	make 1 ochava or dracma....	" ..	55.5 grains,	"
8 ochaves	make 1 onza	" ..	443.8 grains, or 0.9245 ozs. Troy.	
8 onzas	make 1 marco	" ..	3550.5 grains, or 7.396 ozs. Troy.	
2 marcos	make 1 libra	" ..	7101. grains, or 14.7937 ozs. Troy.	

CHINA.

WEIGHTS.

16 taels make 1 catty or pound.............................= 1⅓ lbs. Avoirdupois
100 catties make 1 pecul or tam.............................=133⅓ "   "

CHINESE WEIGHTS INTO TROY.

10 taels	—12.080 ozs.	5 taels	—6.040 ozs.
9 "	—10.872 "	4 "	—4.832 "
8 "	— 9.604 "	3 "	—3.624 "
7 "	— 8.456 "	2 "	—2.416 "
6 "	— 7.248 "	1 "	—1.208 "

MONEY.

10 cash ("le").............make 1 candereen ("fun.")
10 candereens.............make 1 mace ("tseen.")
10 mace.................make 1 tael ("leang.")

BULLION AT HONG KONG OR CANTON.

1 tael	—579.84	grains Troy.
1 tolah	—180	" "
1 tael	— 3.221⅓	tolahs.
717 taels	—1,000	dollars (Mexican.)

97.6 taels Canton weight equal 100 taels Shanghae weight.
1 tael Shanghae weight equals 568 grains Troy.

VALUE OF ONE POUND PURE GOLD AND SILVER AND ONE POUND OF U. S. GOLD AND SILVER COIN.

1 lb. Avoirdupois pure gold is worth $301.56	1 lb. Avoirdupois U.S. gold coin is worth $271.40
1 lb. Troy pure gold is worth.......248.06	1 lb. Troy U. S. gold coin is worth..... 223.25
1 lb. Avoirdupois pure silver is worth.. 18.86	1 lb. Avoirdupois U.S. silver coin is worth 16.97
1 lb. Troy pure silver is worth....... 15.51	1 lb. Troy U. S. silver coin is worth.... 13.96

Table Showing the Proportionate Weight of Gold in a mass of Auriferous Quartz when the Specific Gravity is known.

SPECIFIC GRAVITY.		PROPORTION OF GOLD.	SPECIFIC GRAVITY.		PROPORTION OF GOLD.
2.60	....	0.0000	5.20	....	0.5793
2.65	....	0.0219	5.40	....	0.6007
2.70	....	0.0429	5.60	....	0.6206
2.75	....	0.0632	5.80	....	0.6392
2.80	....	0.0828	6.00	....	0.6565
2.85	....	0.1016	6.20	....	0.6727
2.90	....	0.1198	6.40	....	0.6879
2.95	....	0.1375	6.60	....	0.7021
3.00	....	0.1545	6.80	....	0.7156
3.10	....	0.1869	7.00	....	0.7282
3.20	....	0.2172	7.20	....	0.7402
3.30	....	0.2458	7.40	....	0.7515
3.40	....	0.2726	7.60	....	0.7622
3.50	....	0.2979	7.80	....	0.7724
3.60	....	0.3218	8.00	....	0.7820
3.70	....	0.3444	8.50	....	0.8042
3.80	....	0.3659	9.00	....	0.8239
3.90	....	0.3862	9.50	....	0.8415
4.00	....	0.4055	10.00	....	0.8573
4.10	....	0.4239	10.50	....	0.8717
4.20	....	0.4413	11.00	....	0.8847
4.30	....	0.4580	11.50	....	0.8966
4.40	....	0.4739	12.00	....	0.9075
4.50	....	0.4892	13.00	....	0.9268
4.60	....	0.5037	14.00	....	0.9434
4.70	....	0.5176	15.00	....	0.9577
4.80	....	0.5310	16.00	....	0.9703
4.90	....	0.5438	17.00	....	0.9813
5.00	....	0.5561	18.00	....	0.9912
5.10	....	0.5679	19.00	....	1.0000

SILVER VALUATIONS BASED ON LONDON QUOTATIONS
ADVANCING BY ⅛THS.

London Price. Brit'sh Silver 925 M.	GOLD VALUE. One Ounce 1000 Fine.	One Ounce, U. S. Standard, 900 Fine.	One Dollar in Subsidiary Silver Coin.	$1,000 in Subsidiary Silver Coin.	London Price. Brit'sh Silver 925 M.	GOLD VALUE. One Ounce 1000 Fine.	One Ounce, U. S. Standard, 900 Fine.	One Dollar in Subsidiary Silver Coin.	$1,000 in Subsidiary Silver Coin.
PENCE.	DOLLARS.	DOLLARS.	CENTS.	DOLLARS.	PENCE.	DOLLARS.	DOLLARS.	CENTS.	DOLLARS.
50	1 09.60	98.6,4	79.2,8	792 88.8	53	1 16.18	1 04.5,6	84.0,4	840 46.1
⅛	1 09.88	98.8,9	79.4,8	794 87.0	⅛	1 16.46	1 04.8,1	84.2,4	842 44.4
¼	1 10.15	99.1,4	79.6,8	796 85.2	¼	1 16.73	1 05.0,6	84.4,4	844 42.6
⅜	1 10.43	99.3,8	79.8,8	798 83.5	⅜	1 17.00	1 05.3,0	84.6,4	846 40.8
½	1 10.70	99.6,3	80.0,8	800 81.7	½	1 17.38	1 05.5,5	84.8,3	848 39.0
⅝	1 10.98	99.8,8	80.2,7	802 79.9	⅝	1 17.60	1 05.8,0	85.0,3	850 37.2
¾	1 11.25	1 00.1,2	80.4,7	804 78.1	¾	1 17.82	1 06.0,4	85.2,3	852 35.5
⅞	1 11.52	1 00.3,7	80.6,7	806 76.4	⅞	1 18.09	1 06.2,9	85.4,3	854 33.7
51	1 11.80	1 00.6,2	80.8.7	808 74.6	54	1 18.37	1 06.5,4	85.6,3	856 31.9
⅛	1 12.07	1 00.8,6	81.0,7	810 72.8	⅛	1 18.65	1 06.7,8	85.8,3	858 30.1
¼	1 12.35	1 01.1,1	81.2,7	812 71.0	¼	1 18.92	1 07.0,3	86.0,2	860 28.4
⅜	1 12.62	1 01.3,6	81.4,6	814 69.2	⅜	1 19.29	1 07.2,8	86.2,2	862 26.6
½	1 12.89	1 01.6,0	81.6,6	816 67.5	½	1 19.57	1 07.5,2	86.4,2	864 24.8
⅝	1 13.17	1 01.8,5	81.8,6	818 65.7	⅝	1 19.74	1 07.7,7	86.6,2	866 23.0
¾	1 13.44	1 02.1,0	82.0,6	820 63.9	¾	1 20.02	1 08.0,2	86.8,2	868 21.2
⅞	1 13.72	1 02.3,4	82.2,6	822 62.1	⅞	1 20.39	1 08.2,6	87.0,1	870 19.5
52	1 13.99	1 02.5,9	82.4,6	824 60.4	55	1 20.57	1 08.5,1	87.2,1	872 17.7
⅛	1 14.26	1 02.8,4	82.6,5	826 58.6	⅛	1 20.84	1 08.7,6	87.4,1	874 15.9
¼	1 14.54	1 03.0,8	82.8,5	828 56.8	¼	1 21.11	1 09.0,0	87.6,1	876 14.1
⅜	1 14.81	1 03.3,3	83.0,5	830 55.0	⅜	1 21.39	1 09.2,5	87.8,1	878 12.4
½	1 15.100	1 03.5,8	83.2,5	832 53.2	½	1 21.66	1 09.5,0	88.0,1	880 10.6
⅝	1 15.36	1 03.8,2	83.4,5	834 51.5	⅝	1 21.94	1 09.7,4	88.2,0	882 08.8
¾	1 15.63	1 04.0,7	83.6,4	836 49.7	¾	1 22.21	1 09.9,9	88.4,0	884 07.0
⅞	1 15.90	1 04.3,2	83.8,4	838 47.9	⅞	1 22.48	1 10.2,4	88.6,0	886 05.2

SILVER VALUATIONS BASED ON LONDON QUOTATIONS
ADVANCING BY ⅛THS.

London Price. Brit'sh Silver 925 M.	GOLD VALUE.				London Price. Brit'sh Silver 925 M.	GOLD VALUE.			
	One Ounce 1000 Fine.	Onr Ounce, U. S. Standard, 900 Fine.	One Dollar in Subsidiary Silver Coin.	$1,000 in Subsidiary Silver Coin.		One Ounce 1000 Fine.	One Ounce, U. S. Standard, 900 Fine.	One Dollar in Subsidiary Silver Coin.	$1,000 in Subsidiary Silver Coin.
PENCE.	DOLLARS.	DOLLARS.	CENTS.	DOLLARS.	PENCE.	DOLLARS.	DOLLARS.	CENTS.	DOLLARS.
56	1 22.76	1 10.4,8	88.8,0	888 03,5	58	1 27.14	1 14.4,3	91.9,7	919 75.0
⅛	1 23.03	1 10.7,3	89.0,0	890 01,7	⅛	1 27.42	1 14.6,8	92.1,7	921 73.2
¼	1 23.31	1 10.9,8	89.1,9	891 99,9	¼	1 27.69	1 14.9,2	92.3,7	923 71.5
⅜	1 23.58	1 11.2,2	89.3,9	893 98,1	⅜	1 27.96	1 15.1,7	92.5,6	925 69,7
½	1 23.85	1 11.4,7	89.5,9	895 96,4	½	1 28.24	1 15.4,2	92.7,6	927 67.9
⅝	1 24.13	1 11.7,2	89.7,9	897 94,6	⅝	1 28.51	1 15.6,6	92.9,6	929 66.1
¾	1 24.40	1 11.9,6	89.9,9	899 92,8	¾	1 28.79	1 15.9,1	93.1,6	931 64.4
⅞	1 24.68	1 12.2,1	90.1,9	901 91.0	⅞	1 29.06	1 16.1,6	93.3,6	933 62.6
57	1 24.95	1 12.4,6	90.3,8	903 89.2	59	1 29.33	1 16.4,0	93.5,6	935 60.8
⅛	1 25.22	1 12.7,0	90.5,8	905 87.5	⅛	1 29.61	1 16.6,5	93.7,5	937 59.0
¼	1 25.50	1 12.9,5	90.7,8	907 85.7	¼	1 29.88	1 16.9,0	93.9,5	939 57.2
⅜	1 25.77	1 13.2,0	90.9,8	909 83.9	⅜	1 30.16	1 17.1,4	94.1,5	941 55.5
½	1 26.05	1 13.4,4	91.1,8	911 82.1	½	1 30.43	1 17.3,9	94.3,5	943 53.7
⅝	1 26.32	1 13.6,9	91.3,8	913 80.4	⅝	1 30.70	1 17.6,3	94.5,5	945 51.9
¾	1 26.59	1 13.9,4	91.5,7	915 78.6	¾	1 30.98	1 17.8,8	94.7,5	947 50.1
⅞	1 26.87	1 14.1,8	91.7,7	917 86.8	⅞	1 31.25	1 18.1,3	94.9,4	949 48.4
					60	1 31.53	1 18.3,7	95.1,4	951 46.6

Method of Determining the respective quantities of Gold and Quartz in Quartz Specimens.

$$\frac{\text{S. G. Nugget}-\text{S. G. Quartz}}{\text{S. G. Gold}-\text{S. G. Quartz}} \times \frac{\text{S. G. Gold}}{\text{S. G. Specimen}} \times \text{Weight of Nugget} = \text{Weight of Gold in Specimen.}$$

$$\frac{\text{S. G. Gold}-\text{S. G. Specimen}}{\text{S. G. Gold}-\text{S. G. Quartz}} \times \frac{\text{S. G. Quartz}}{\text{S. G. Specimen}} \times \text{Weight of Specimen} = \text{Weight of Quartz in Specimen.}$$

1st EXAMPLE—Suppose we have a specimen weighing 1000 grains, and its specific gravity is found to be 8—Suppose, also, the specific gravity of gold were 18, and that of quartz 2, then, by above rule:

$$\frac{8-2}{18-2} \times \frac{18}{8} \times 1000 = \frac{6}{16} \times \frac{18}{8} \times 1000 = 843\tfrac{3}{4} \text{ grains of gold in Specimen.}$$

$$\frac{18-8}{18-2} \times \frac{2}{8} \times 1000 = \frac{10}{16} \times \frac{2}{8} \times 1000 = 156\tfrac{1}{4} \text{ quartz in Specimen.}$$

2d EXAMPLE—Nugget weighing 10 oz., S. G. 8½; S. G. gold 19. S. G. quartz 2½:—

$$\frac{8.5-2.5}{19.0-2.5} \times \frac{19}{8.5} \times 10 = \frac{6}{16\frac{1}{2}} \times \frac{19}{8\frac{1}{2}} \times 10 = 8.1283 \text{ of gold in Specimen.}$$

$$\frac{10.00-8.5}{19.0-2.5} \times \frac{2.5}{8.5} \times 10 = \frac{10\frac{1}{2}}{16\frac{1}{2}} \times \frac{2\frac{1}{2}}{8\frac{1}{2}} \times 10 = 1.8716 \text{ quartz in Specimen.}$$

RELATIVE VALUE OF AVOIRDUPOIS AND TROY WEIGHTS.

Avoirdupois Ounces.	Troy. Ozs.	Troy. Dwts.	Troy. Grs.	Grains.	Troy. Ounces.	Avoirdupois. Ozs.	Avoirdupois. Drachms.	Grains.
.... 1..		18..	5½..	.... 437½ ..	.... 1...	.... 1...	.. 2..	... 480...
.... 2..	.... 1.	....16.	...11 ...	... 875 ..	.... 2...	.... 2...	... 3...	.... 960...
.... 3..	.... 2.	....14.	...16½...	...1312½..	.... 3...	.... 3...	... 5...	...1440...
.... 4..	.... 3.	....12.	...22 ...	...1750 ..	.... 4...	.... 4...	... 6...	...1920...
.... 5..	.... 4.	....11.	3½...	...2187½..	.... 5...	.... 5...	... 8...	...2400...
.... 6..	.... 5.	.... 9.	... 9 ..	...2625 ..	.... 6...	.... 6...	... 9...	...2880...
.... 7..	.... 6.	.... 7.	...14½..	...3062½..	.... 7...	.... 7...	...11...	...3360...
.... 8..	.... 7.	.... 5.	...20...	...3500 ..	.... 8...	.... 8...	...12...	...3840...
.... 9..	.... 8.	.... 4.	1½...	...3937½..	.... 9...	.... 9...	...14...	...4320...
...10..	.... 9.	.... 2.	... 7 ..	...4375 ..	...10...	...11...	.... 0...	...4800...
...11..	...10.	.... 0.	...12½..	...4812½..	...11...	...12...	... 1...	...5280...
...12..	...10.	...18.	...18 ..	...5250 ..	...12...	...13...	... 3...	...5760...
...13..	...11.	...16.	...23½..	...5687½..				
...14..	...12.	...15.	... 5 ..	...6125 ..				
...15..	...13.	...13.	...10½..	...6562½..				
...16..	...14.	...11.	...16 ..	...7000 ..				

AVOIRDUPOIS.

		Drachms.	Ounces.	Pounds.
16 drachms.............	=1 ounce.			
16 ounces.............	=1 pound......=	256		
112 pounds.............	=1 cwt........=	28672......=	1792	
20 cwt	=1 ton ....=	573440......=	35840.......=	2240

TROY.

		Grains.	Dwts.
24 grains............	=1 dwt.		
20 dwts.............	=1 ounce......=	480	
12 ounces	=1 pound.......=	5760......=	240

www.ingramcontent.com/pod-product-compliance
Lightning Source LLC
Chambersburg PA
CBHW031803090426
42739CB00008B/1139